Portfolio Rebalancing

Financial Mathematics Series
Aims and scope

The field of financial mathematics forms an ever-expanding slice of the financial sector. This series aims to capture new developments and summarize what is known over the whole spectrum of this field. It will include a broad range of textbooks, reference works and handbooks that are meant to appeal to both academics and practitioners. The inclusion of numerical code and concrete real-world examples is highly encouraged.

Series Editors

M.A.H. Dempster
Centre for Financial Research
Department of Pure Mathematics and Statistics
University of Cambridge

Dilip B. Madan
Robert H. Smith School of Business
University of Maryland

Rama Cont
Department of Mathematics
Imperial College

Stochastic Finance
A Numeraire Approach
Jan Vecer

Equity-Linked Life Insurance
Partial Hedging Methods
Alexander Melnikov, Amir Nosrati

High-Performance Computing in Finance
Problems, Methods, and Solutions
M. A. H. Dempster, Juho Kanniainen, John Keane, Erik Vynckier

Derivative Pricing
A Problem-Based Primer
Ambrose Lo

An Introduction to Computational Risk Management of Equity-Linked Insurance
Runhuan Feng

Portfolio Rebalancing
Edward E. Qian

For more information about this series please visit: https://www.crcpress.com/Chapman-and-HallCRC-Financial-Mathematics-Series/book-series/CHFINANCMTH

Portfolio Rebalancing

By
Edward E. Qian

CRC Press is an imprint of the
Taylor & Francis Group, an **informa** business
A CHAPMAN & HALL BOOK

CRC Press
Taylor & Francis Group
6000 Broken Sound Parkway NW, Suite 300
Boca Raton, FL 33487-2742

First issued in paperback 2020

© 2019 by Taylor & Francis Group, LLC
CRC Press is an imprint of Taylor & Francis Group, an Informa business

No claim to original U.S. Government works

ISBN-13: 978-1-4987-3244-4 (hbk)
ISBN-13: 978-0-367-73283-7 (pbk)

Library of Congress Cataloging-in-Publication Data

Names: Qian, Edward E., author.
Title: Portfolio rebalancing / Edward E. Qian.
Description: Boca Raton, Fla. : CRC Press, [2019] | Series: Financial mathematics series
Identifiers: LCCN 2018030136| ISBN 9781498732444 (hardback : alk. paper) | ISBN 9781315120676 (ebook)
Subjects: LCSH: Portfolio management--Mathematical models.
Classification: LCC HG4529.5 .Q249 2019 | DDC 332.6--dc23
LC record available at https://lccn.loc.gov/2018030136

Visit the Taylor & Francis Web site at
http://www.taylorandfrancis.com

and the CRC Press Web site at
http://www.crcpress.com

For My Family

Contents

Preface

PORTFOLIO REBALANCING IS A standard practice in managing many types of investment portfolios. For example, institutional investors such as pension funds and sovereign wealth funds regularly rebalance their investment portfolios back to long-term strategic asset allocation policy. Leveraged exchange-traded-funds (ETF), popular among short-term investors, carry out portfolio rebalancing on a daily basis to maintain funds' stated leverage ratios to the underlying indices. What are the purposes and possible benefits of portfolio rebalancing as opposed to buy-and-hold?

One purpose is risk management. Without portfolio rebalancing, a buy-and-hold portfolio can drift away from its predetermined benchmark, resulting in a large tracking error. In addition, the portfolio on its own can become non-diversified over time. Therefore, a rebalanced portfolio in general has lower risk than its buy-and-hold counterpart. A natural question follows: does a rebalanced portfolio also have a higher return than the buy-and-hold portfolio? In other words, is there a rebalancing alpha?

This book has grown mostly out of the author's search for answers to this question. The answer appears to be yes, based on the notion of diversification return, which is almost always positive for a long-only portfolio. However, this claim is false, since diversification return is, in general, not the return difference between the rebalanced portfolio, or fixed-weight portfolio, and the buy-and-hold portfolio. Diversification return is only one part of the return difference, stemming from the volatilities of investment returns. The other part comes from differences between investment returns, which are always negative for a long-only portfolio. A rebalancing alpha thus depends on the relative magnitude of the two parts.

Portfolio rebalancing of long-only portfolios involves buying winners and selling losers—a contrarian investment strategy. A crucial question is: does a positive rebalancing alpha necessitate mean reversion in asset returns? The answer to this question requires a direct comparison between the cumulative wealth of the fixed-weight portfolio and the buy-and-hold portfolio instead of analysis based on approximations. Nowadays, there are various alternative investment strategies that are long-short, for which portfolio rebalancing has a different mechanism. The answer to the question of rebalancing alpha for a long-short portfolio is more difficult to find, and it can be quite different from the answer for a long-only portfolio.

This book aims to summarize the search and answers, both complete and incomplete, for rebalancing alpha in a self-contained volume, using both quantitative and empirical analysis. The quantitative analysis relies on classical methods of mathematics rather than stochastic calculus. Therefore, it is accessible to readers who have knowledge of univariate calculus and undergraduate statistics. The empirical analysis covers asset allocation portfolios as well as portfolios of stocks, bonds, and commodities. Consistently with analytic results, the empirical results show that rebalancing alpha can be either positive or negative. Rebalancing alpha does not always exist.

In recent years, there have been many advances in the analysis of rebalancing alpha since the early introduction of diversification return by Booth and Fama (1992). For example, Hallerbach (2014) outlines a framework of analyzing rebalancing alpha that is similar to the one used in the book. Rabault (2017) presents a connection between rebalancing alpha and mean reversion of returns. A list of research papers is included in the bibliography. It is noted that the material selected for the book reflects the perspectives as well as the limitations of the author. It is not intended as a comprehensive survey of this interesting topic, and omissions are regrettable. All errors are the author's own.

CHAPTER 1

Introduction

PORTFOLIO REBALANCING IS AN essential part of portfolio management. Although portfolio rebalancing can be applied to any portfolios, such as asset allocation, equity, fixed income, commodity, and
hedge fund portfolios, its earliest application was in strategic asset allocation portfolios. These multi-asset portfolios are chosen by investors to
meet their long-term return objective and risk tolerance after taking into
account expected returns and risks of underlying asset classes. Investors
are expected to adhere to the strategic asset allocation, which also serves
as a performance benchmark over the appropriate investment horizon.

1.1 RISK MANAGEMENT

The main objective of portfolio rebalancing, whether it is done on a calendar basis or with a threshold, is risk management. Specifically, the aim
of portfolio rebalancing is to maintain portfolio risk within risk tolerance.
Without portfolio rebalancing, differences in asset returns can cause an
asset allocation portfolio to drift toward a portfolio that has a different
risk profile in terms of total portfolio risk and risk contributions. Take, for
example, a 60/40 portfolio with 60% of capital invested in stocks and 40%
of capital invested in bonds. Stocks have higher risks than bonds and thus
tend to have higher expected and often realized returns than bonds. As
a result, a passive approach, which is often referred to as a *buy-and-hold
approach,* would lead the weight in stocks to drift higher and the weight
in bonds to drift lower. As a simple numerical illustration, suppose stocks
and bonds have annualized returns of 10% and 5%, respectively, during a
five-year period. Left alone for five years, a 60/40 portfolio at the beginning

of the period will be close to a 65/35 portfolio. Because of higher allocation to stocks—a higher-risk asset—the 65/35 portfolio has higher risk than the original 60/40 portfolio. Furthermore, a 5% overweight in stocks at the expense of bonds exposes the portfolio to additional risks in stocks. Of course, the bias toward stocks caused by portfolio drift can be beneficial to portfolio returns, since more often than not, stocks tend to outperform bonds. In a normal market environment, higher risks usually lead to higher returns. In fact, many asset managers intentionally embed this bias into their investment process and then dubiously claim the excess return as alpha, which is actually reward for additional risk.

Of course, this additional risk in stocks can also be detrimental at times. When a bear market or market crash occurs, the 5% overweight in stocks can cause additional drawdown for asset allocation portfolios. Imagine that at the beginning of 2008, we have two portfolios: one is the 60/40 portfolio that had been rebalanced, and the other is the 65/35 portfolio. During the year 2008, stocks lost about 40% of their value as measured by the Morgan Stanley Capital International (MSCI) world index, while bonds gained about 10% as measured by the World Government Bond Index (WGBI). For the year 2008, the 60/40 portfolio was down 20%, and the 65/35 portfolio was down 22.7% with an additional loss of 2.7%. Indeed, even though in this case a buy-and-hold approach might lead to higher return over the long term relative to a rebalanced portfolio by taking on more risks, it might suffer negative excess returns in any given period. A systematic portfolio rebalancing approach can help mitigate those losses.

1.2 REBALANCING ALPHA

Does portfolio rebalancing necessarily lead to a return advantage compared with a buy-and-hold approach, as we indicated in the last section? Can portfolio rebalancing generate a rebalancing alpha? In other words, can a rebalanced portfolio with lower risk outperform a buy-and-hold portfolio with higher risk? To some investors, this rebalancing alpha is another objective of portfolio rebalancing. There have been cases in which some investments have been able to generate higher returns with lower risks versus standard indices. Examples include value stock and low-volatility stocks versus capitalization-weighted indices.

The previous example hinted at this possibility. At the end of 2008, the 60/40 portfolio became a 45/55 portfolio with 45% in stocks and 55% in bonds because of dramatic underperformance of stocks. Meanwhile, the 65/35 portfolio became a 51/49 portfolio. What would happen if we rebalanced the first portfolio to 60/40 again and did not rebalance the second portfolio? It turns out that stocks rebounded strongly in 2009, with the MSCI world index returning about 23% and the WGBI index returning about 1%. The 60/40 portfolio saw a return of 14.2%, while the 51/49 portfolio saw a return of 12.2%. Hence, the rebalanced portfolio outperformed the buy-and-hold portfolio again in 2009 by 2%. However, the fact that the rebalanced 60/40 portfolio performed better than the buy-and-hold portfolio during the years 2008 and 2009 is a special case. First, the capital market was in turmoil during and after the global financial crisis. Second, the investment horizon of two years is simply too short to prove the existence of rebalancing alpha.

To answer the question of whether portfolio rebalancing adds value or not, we need more research in terms of both empirical study and theoretical investigation. Unfortunately, no empirical study of portfolio rebalancing has produced any concrete answer to the question. In fact, many empirical studies have added more confusion to the topic, because some studies show that portfolio rebalancing added value, while other studies show just the opposite. This is not unexpected, since any empirical study is specific to asset classes, portfolio weights, and characteristics of asset returns during the time periods chosen. General claims about rebalancing alpha or the lack of it based on one specific empirical study cannot be substantiated. The results of empirical studies raise a further question, however: why portfolio rebalancing adds value in some cases but not in other cases. Perhaps these questions can be tackled by theoretical research on portfolio rebalancing.

1.3 DIVERSIFICATION RETURN, VOLATILITY EFFECT

One of the important and unfortunately often misunderstood theoretical concepts in portfolio rebalancing is diversification return. Diversification return refers to the difference between the geometric return of a fixed-weight portfolio (i.e. a rebalanced portfolio) and the weighted average of geometric returns of the underlying investments. Take the 60/40 portfolio as an example. The diversification return equals the geometric return of a rebalanced 60/40 portfolio minus the weighted sum of the geometric returns of stocks and bonds, with the weights being 60% for stocks and 40% for bonds.

Mathematically, it can be proved that for a long-only portfolio, diversification return is always non-negative. The reason why it is called *diversification return* has to do with portfolio diversification of risk, since the mathematical proof hinges on the fact that the variance of a fixed-weight portfolio is always less than or equal to the weighted average of variances of underlying investments. The term *diversification* is not the source of misunderstanding.

The misunderstanding of diversification return is probably caused by the term *return*. Some investors and researchers falsely think that diversification return is the return difference between the geometric return of the rebalanced portfolio and the geometric return of the corresponding buy-and-hold portfolio. Since diversification return is always non-negative for a long-only portfolio, many are led to believe that diversification return is rebalancing alpha. In other words, portfolio rebalancing always adds value for a long-only portfolio.

This notion is wrong, because the weighted average of geometric returns of underlying investments is not the geometric return of the buy-and-hold portfolio. In fact, it is generally not the return of any portfolio. In some sense, it is a mathematical construct based on underlying returns, to which the fixed-weight portfolio is compared. It turns out that this comparison is analytically tractable through the variance of the fixed-weight portfolio and the variances of underlying investments. Nevertheless, diversification return is not rebalancing alpha, because it is not the return difference between the rebalanced portfolio and the buy-and-hold portfolio.

It is in fact *half* of rebalancing alpha. To complete the analysis, the next logical step is to compare the geometric return of the buy-and-hold portfolio with the same weighted average of underlying returns. It can be shown that for a long-only portfolio, the geometric return of a buy-and-hold portfolio is also always greater than or equal to the weighted average of the geometric returns of underlying components. If we denote the geometric return of the fixed-weight portfolio by g_{FW}, that of the buy-and-hold portfolio by g_{BH}, and the weighted average of geometric returns of underlying investments by $\bar{g}$, we have $g_{BH} \geq \bar{g}$ and $g_{FW} \geq \bar{g}$. The triangulation of three terms can only be completed by comparing the two differences: $g_{BH} - \bar{g}$ versus $g_{FW} - \bar{g}$. We call them the *return effect* and the *volatility effect* of portfolio rebalancing, respectively. If the return effect is less than the volatility effect, rebalancing alpha is positive. But if the return effect is greater than the volatility effect, rebalancing alpha would be negative. The establishment of this framework allows a thorough analysis of both effects to determine the overall rebalancing alpha.

1.4 SERIAL CORRELATION AND REBALANCING ALPHA

It is natural to suspect that rebalancing alpha is related to serial correlations of asset returns. Specifically, positive serial correlations or return momentum lead to negative rebalancing alpha, while negative serial correlations or return reversal lead to positive rebalancing alpha. This is supported by our example of the 60/40 portfolio during the year 2008 and 2009. In 2008, stocks underperformed bonds significantly, and the pattern reversed in 2009, when stocks outperformed bond significantly. There was a strong reversal of relative return between stocks and bonds. The rebalance at the end of 2008 of buying stocks (previous loser) and selling bonds (previous winner) helped the portfolio in 2009, when winner and loser reversed. If, however, stocks had underperformed bonds again in 2009, or if there had been a trend in the relative return between the two assets, the rebalance at the end of 2008 would have had a negative impact on the portfolio.

So, does positive rebalancing alpha necessitate mean reversion in asset returns? This question is frequently debated among practitioners and academics. The answer seems to be yes, at least from an academic perspective. It can be proved statistically that if underlying returns are serially uncorrelated and have the same expected value, then the expected returns of a fixed-weight portfolio and a buy-and-hold portfolio would be identical; that is, no rebalancing alpha.

This is a very clear and strong result in theory. In practice, however, no returns fit these statistical assumptions perfectly. What can be expected of serial correlations of returns when rebalancing alpha is either positive or negative? The analysis of volatility and return effects with approximations seems inadequate to answer this question. However, an exact treatment of rebalancing alpha can be used for studying this question. We shall provide evidence that a positive rebalancing alpha does require some degree of mean reversion.

The structure of serial correlations of different lags can also have an impact on rebalancing alpha of varying frequency and/or thresholds. For example, if asset returns exhibit short-term momentum and long-term reversal, it is probably optimal to carry out portfolio rebalancing with a longer horizon or with a larger threshold. In this area of research, one often has to use numerical simulations to analyze different options.

1.5 NEW TOPICS IN PORTFOLIO REBALANCING

Most previous research on portfolio rebalancing focuses on long-only portfolios. This is quite understandable, because most investors in the past invested in traditional long-only portfolios without leverage. In recent years, there has been growing development and acceptance by investors of more sophisticated investment strategies that use leverage and have both long and short positions. Examples include

- Risk parity multi-asset portfolios

- Hedge fund strategies

- Equity 130/30 strategies

- Leveraged and short exchange-traded-funds (ETFs)

All these strategies can be either long-only or long-short, but they all use leverage.

New dynamics emerge when portfolio rebalancing is applied to these portfolios. For instance, portfolio rebalancing of long-short portfolios might be implemented with buying winners and selling losers instead of buying losers and selling winners as in long-only portfolios. Diversification return can be negative for a long-short portfolio, while it is always positive for long-only portfolios. For a portfolio with leverage, the degree of leverage changes over time due to profit or loss of the portfolio. Portfolio rebalancing also involves a rebalancing of leverage. This is not an issue for long-only portfolios, whose leverage remains constant. New research is needed to understand how portfolio rebalancing works and how it affects the investment performance of these portfolios. For example, we need to understand why levered ETFs (ultra-long or ultra-short) might not deliver returns close to their stated multiples of the underlying index returns.

Investment performance should be about not only return or alpha but also risk. The correct measure is risk-adjusted return, or the Sharpe ratio. In many cases, portfolio rebalancing may lead to a lower return than a buy-and-hold approach. But it also results in lower risk. The risk-adjusted return of fixed-weight portfolios versus buy-and-hold portfolios is a new research area.

1.6 OUTLINE OF THE BOOK

The remaining two chapters of Part I of the book introduce basic concepts used in the book. Chapter 2 defines portfolio returns (both arithmetic and geometric), return volatility, and serial correlation. It also presents relationships between arithmetic and geometric return, which will be used to derive diversification return. Chapter 3 introduces portfolio rebalancing for both long-only and long-short portfolios and defines rebalancing alpha. We present examples of portfolio rebalancing with positive and negative rebalancing alphas.

Part II of the book presents analysis of rebalancing alpha using the framework of volatility effect and return effect. The definition is given in Chapter 4, followed by the analysis of volatility effect in Chapter 5, of return effect in Chapter 6, and of the combined effect in Chapter 7.

Part III performs empirical analysis of portfolio rebalancing with the framework established in Part II. We examine traditional asset allocation portfolios as well as risk parity multi-asset portfolios in Chapter 8. With regard to underlying asset classes, we examine in Chapter 9 portfolio rebalancing of equity, fixed income, and commodity portfolios. We relate rebalancing alpha of these portfolios to volatility and return effects.

Part IV covers more recent research topics. Chapter 10 presents an exact formula of rebalancing alpha for any arbitrary rebalancing schedule. We show why positive rebalancing alpha requires some degree of mean reversion. Chapter 11 analyzes the risk-adjusted return (or wealth) of both fixed-weight portfolios and buy-and-hold portfolios. Finally, in Chapter 12, we study threshold rebalancing using numerical simulation.

CHAPTER 2

A Brief Review of Portfolio Theory

IN THIS CHAPTER, WE introduce mathematical notations that will be used throughout the book and provide a brief review of portfolio theory that is relevant to the analysis of portfolio rebalancing. For the most part, our analysis does not require distributional assumption of underlying security returns. In some of the later chapters where a return distribution is required, we use multivariate normal distribution to facilitate the analysis.

Let us consider for the moment that there is one investment asset and an investment horizon of N periods. We denote the return in period j by $r_j, j = 1, \ldots, N$. The return has no theoretical upper bound, and for limited liability investments, it is bounded below by -1, which might not be true for investments with leverage. We now consider arithmetic and geometric means of returns over N periods.

2.1 ARITHMETIC AND GEOMETRIC MEANS

The arithmetic mean or average return is defined as

$$\mu = \frac{1}{N} \sum_{j=1}^{N} r_j = \frac{1}{N} \left(r_1 + \cdots + r_N \right) \qquad (2.1)$$

The geometric mean, on the other hand, is defined as

$$g = \sqrt[N]{\prod_{j=1}^{N}(1+r_j)} - 1 = \left[(1+r_1)\cdots(1+r_N)\right]^{1/N} - 1 \qquad (2.2)$$

It is usually more natural to write Equation 2.2 in the following form, which shows that g is the average cumulative return of the asset over N periods:

$$(1+g)^N = \prod_{j=1}^{N}(1+r_j) = (1+r_1)\cdots(1+r_N) \qquad (2.3)$$

One elementary result concerning arithmetic and geometric means is that the former is always greater than or equal to the latter. Before proving this fact mathematically, we provide a simple yet insightful example.

Example 2.1: Suppose an investment has a return of 100% in the first period followed by a return of −50% in the second period. Then, the arithmetic return is

$$\mu = \frac{1}{2}(1-0.5) = 25\%$$

On the other hand, the geometric mean is

$$g = \sqrt[2]{(1+1)(1-0.5)} - 1 = 0\%$$

The geometric mean is zero, because the value of the investment has not changed. It doubled in the first period and then halved in the second period, with a cumulative return of 0%. However, the arithmetic mean is quite positive. Intuitively, if an investment has zero cumulative return after two periods, the returns from the two periods must be of opposite signs, In addition, the magnitude of the positive return must be higher than that of the negative one, leading to a positive arithmetic mean.

To prove $\mu \geq g$ mathematically, we first note that when $r_j \geq -1$, then $1 + r_j \geq 0$. The following arithmetic mean–geometric mean (AM-GM) inequality holds for general non-negative numbers $a_1, a_2, \cdots, a_N$:

$$\frac{a_1 + a_2 + \cdots + a_N}{N} \geq \left(a_1 a_2 \cdots a_N \right)^{1/N} \tag{2.4}$$

Now, we substitute $1 + r_j$ for a_j with $j = 1, \cdots, N$. The left-hand side of Equation 2.4 becomes $1 + \mu$, and the right side becomes $1 + g$. It follows that $\mu \geq g$. We have the following remarks about the result:

- The condition $1 + r_j \geq 0$ must be true for the AM-GM inequality to hold.
- The equality holds only when returns from all the periods are identical, i.e. $r_1 = r_2 = \cdots = r_N$. This implies that the volatility of returns of the asset is zero. We shall analyze the relationship between arithmetic and geometric means and return volatility in more detail.

The AM-GM inequality (Equation 2.4) is a special case of a more general result. Note that the left-hand side represents an equally weighted average. If we have a weighted average with non-negative weights $p_1, p_2, \ldots, p_N$ and $p_1 + p_2 + \cdots + p_N = 1$, and $a_1, a_2, \ldots, a_N$ are non-negative real numbers, the generalized AM-GM inequality is

$$p_1 a_1 + p_2 a_2 + \cdots + p_N a_N \geq a_1^{p_1} a_2^{p_2} \cdots a_N^{p_N} \tag{2.5}$$

- For Equation 2.5 to be an equality, all $a_1, a_2, \ldots, a_N$ must be the same. We shall use this inequality when we study arithmetic and geometric returns of portfolios of assets.

2.2 RETURN VOLATILITIES

For individual assets, return volatility is the standard deviation of returns over multiple periods. We define it through variance of the returns:

$$\sigma^2 = \frac{1}{N} \left[\left(r_1 - \mu \right)^2 + \left(r_2 - \mu \right)^2 + \cdots + \left(r_N - \mu \right)^2 \right] \tag{2.6}$$

The variance can also be written in terms of the arithmetic mean, or the first moment, and the second moment. We have

$$\sigma^2 = \frac{1}{N}\left(r_1^2 + r_2^2 + \cdots + r_N^2\right) - \mu^2 \triangleq \overline{r^2} - \left(\overline{r}\right)^2 \tag{2.7}$$

- where the bars denote averages. Note that Equations 2.6 and 2.7 are defined with N as the divisor instead of $N-1$. As a result, it is not intended as the unbiased estimator of variance. Our goal is to analyze the relationship between variance and various returns.

Example 2.2: From Example 2.1, we can calculate the return volatility of an investment that has a return of 100% and a return of −50%. We have

$$\sigma = \sqrt{\frac{1}{2}\left[\left(1-0.25\right)^2 + \left(-0.5-0.25\right)^2\right]} = 75\%$$

The variance (Equation 2.6) is defined in terms of differences between individual returns and the arithmetic average. It can be rewritten as pairwise differences among individual returns. We have

$$\sigma^2 = \frac{1}{2N^2} \sum_{i,j=1}^{N}\left(r_i - r_j\right)^2 \tag{2.8}$$

The double summation is over all possible i and j. Of course, when i and j are the same, the term vanishes. The proof of Equation 2.8 is left as an exercise.

We have defined the arithmetic mean and the return volatility using equal weight, namely $1/N$ for all the returns. The definition can be generalized to unequal weights. If the weights are $p_1, p_2, \cdots, p_N$ for $r_1, r_2, \cdots, r_N$, respectively, with $p_1 + p_2 + \cdots + p_N = 1$, we define

$$\mu = p_1 r_1 + p_2 r_2 + \cdots + p_N r_N \tag{2.9}$$

Similarly,

$$\sigma^2 = p_1\left(r_1 - \mu\right)^2 + p_2\left(r_2 - \mu\right)^2 + \cdots + p_N\left(r_N - \mu\right)^2 \tag{2.10}$$

Note that the mean μ is given by Equation 2.9.

- The definitions 2.9 and 2.10 are used when one wants to assign different weights to returns from different periods. A typical method is to give more weight to recent data points and less weight to past data points using an exponentially decayed weighting scheme.
- In terms of algebraic expressions, the weights $p_1, p_2, \cdots, p_N$ in Equations 2.9 and 2.10 don't have to be positive. When we have negative weights, both definitions still stand, but their interpretation might have to change. For instance, the term in Equation 2.10 may no longer be non-negative—it is not the traditional variance we are accustomed to. We shall encounter such a generalized "variance" later in the book when we study long-short portfolios that have both positive and negative portfolio weights.

The generalized variance (Equation 2.10) can be written in a form that is similar to Equation 2.8. We have

$$\sigma^2 = \frac{1}{2} \sum_{i,j=1}^{N} p_i p_j \left(r_i - r_j \right)^2 \tag{2.11}$$

2.3 RELATIONSHIPS BETWEEN ARITHMETIC AND GEOMETRIC MEANS

The AM-GM inequality shows that the arithmetic mean is in general higher than the geometric mean. The question we now turn to is "by how much?" We derive an approximate relationship between the two means, which will serve as a useful tool in the study of portfolio rebalancing.

2.3.1 Analytic Approximation

Taking the logarithm of Equation 2.3, we have

$$\ln\left(1+g\right) = \frac{1}{N}\left[\ln\left(1+r_1\right) + \cdots + \ln\left(1+r_N\right)\right] \tag{2.12}$$

When the returns are small, we can approximate the logarithmic functions on the right by a Taylor expansion around zero, i.e. $\ln\left(1+r_i\right) \approx r_i - r_i^2/2 + O\left(r_i^3\right)$. The error of the approximation is of the third power of individual returns. Since we aim to relate the geometric mean g to the arithmetic mean μ, a more direct method is to expand the logarithmic functions around μ:

$$\ln(1+r_i)=\ln(1+\mu)+\frac{r_i-\mu}{1+\mu}-\frac{(r_i-\mu)^2}{2(1+\mu)^2}+\cdots+(-1)^{k-1}\frac{(r_i-\mu)^k}{k(1+\mu)^k}+\cdots \tag{2.13}$$

This series converges when

$$-1<\frac{r_i-\mu}{1+\mu}\leq 1 \text{ or } -1<r_i\leq 1+2\mu. \tag{2.14}$$

Condition 2.14 should be satisfied when the individual returns are modest. However, an extremely positive outliner, much higher than the rest of the returns, might not meet the condition.

Substituting Equation 2.13 into Equation 2.12 leads to

$$\ln(1+g)=\ln(1+\mu)-\frac{\sigma^2}{2(1+\mu)^2}+\cdots+(-1)^{k-1}\frac{s_k}{k(1+\mu)^k}+\cdots \tag{2.15}$$

The term s_k is simply

$$s_k=\frac{1}{N}\sum_{i=1}^{N}(r_i-\mu)^k$$

For example, s_2 is variance, s_3 is skewness multiplied by volatility cubed, s_4 is kurtosis multiplied by variance squared, and so on. Taking the exponential function of Equation 2.15, we obtain

$$1+g=(1+\mu)\exp\left[\sum_{k=2}^{\infty}\frac{(-1)^{k-1}s_k}{k(1+\mu)^k}\right] \tag{2.16}$$

If we just take the first term in the summation, we have

$$1+g\approx(1+\mu)\exp\left[-\frac{\sigma^2}{2(1+\mu)^2}\right] \tag{2.17}$$

The approximation can be further simplified by approximating the exponential term by $\exp(x) \approx 1 + x$ and discarding higher-order terms in the resulting multiplication. We obtain

$$g \approx \mu - \frac{\sigma^2}{2} \qquad (2.18)$$

Equation 2.18 is a linear relationship between the arithmetic and geometric means and the return variance. This makes many studies, including portfolio rebalancing, analytically tractable. It is also reasonably accurate for many types of investment returns. For example, Mindlin (2011) examined the accuracy of four different approximations of geometric means with varying complexity and found that even though results given by Equation 2.18 are often less accurate than results from more sophisticated equations, they are accurate enough for many practical purposes.

- The approximation error of Equation 2.18 relative to Equation 2.17 is in the order of $O(\mu\sigma^2)$. When both μ and σ are small, it is likely that the error of Equation 2.17 relative to Equation 2.16 will be smaller than $O(\mu\sigma^2)$. Under these conditions, the error of using $\mu - \sigma^2/2$ for the approximation of geometric return is in the order of $O(\mu\sigma^2)$.

Example 2.3: From Examples 2.1 and 2.2, we have $\mu = 25\%$ and $\sigma = 75\%$. Hence,

$$g \approx \mu - \frac{\sigma^2}{2} = 0.25 - \frac{0.75^2}{2} = -3.13\%$$

The true geometric return is 0%. The approximation underestimates the true value by 3.13%.

2.3.2 Empirical Examination

It is hard to know from Example 2.3 whether the approximation (Equation 2.18) is accurate or not. We now apply it to actual historical returns of different asset classes, including cash, bonds, stocks, and commodities. The indices for these asset classes are three-month U.S. Treasury bills, 10-year U.S. Treasury bonds, the S&P 500 index, and the Goldman Sachs Commodity Index (GSCI), and their annual returns span from 1970 to 2014.

TABLE 2.1   Statistics of Asset Returns from 1970 to 2014

	Cash	Bonds	Stocks	Commodities
g	5.16%	7.58%	10.48%	8.03%
μ	5.21%	7.84%	11.97%	11.02%
σ	3.38%	7.60%	17.24%	24.66%
g'	5.15%	7.55%	10.48%	7.98%
e	0.00%	−0.03%	0.00%	−0.05%
e_r	6×10^{-4}	4×10^{-3}	2×10^{-4}	6×10^{-3}

Table 2.1 lists the return statistics of their returns over the whole sample. In the table, $g' = \mu - \sigma^2/2$, the error term $e = g' - g$, and finally, $e_r = |e/g|$ is the relative error. We first note that the returns of all asset classes are quite positive, with the S&P index having the highest return. In terms of volatility, stocks and commodities have significant risks. On a risk-adjusted basis, commodities are not as attractive as other assets. We shall formally define the Sharpe ratio later.

The table shows that $g' = \mu - \sigma^2/2$ approximates g extremely well for all asset classes. The difference between the two is roughly zero for cash and stocks. It is only a few basis points (bps; one basis point is 0.01%) for bonds and commodities.

The accuracy of the approximations declines slightly over sub-sample periods. We break the period of 45 years into two sub-periods: one from 1970 to 1992 and the other from 1993 to 2014. The return statistics are listed in Tables 2.2 and 2.3, respectively. In both periods, the approximation of geometric return for commodities worsened, with an error of −35 bps in the first period and −17 bps in the second period. The approximation error for stocks in the second period is also relatively high at −19 bps. It is also noted that all returns are lower in the second period than in the first period. The return of commodities is especially weak in the second period, being only 1.38%.

TABLE 2.2   Statistics of Asset Returns from 1970 to 1992

	Cash	Bonds	Stocks	Commodities
g	7.50%	9.48%	11.46%	14.80%
μ	7.53%	9.74%	12.70%	16.79%
σ	2.65%	7.73%	16.07%	21.60%
g'	7.49%	9.44%	11.41%	14.45%
e	0.00%	−0.04%	−0.05%	−0.35%
e_r	4×10^{-4}	4×10^{-3}	5×10^{-3}	2×10^{-2}

TABLE 2.3   Statistics of Asset Returns from 1993 to 2014

	Cash	Bonds	Stocks	Commodities
g	2.77%	5.63%	9.46%	1.38%
μ	2.79%	5.86%	11.20%	5.00%
σ	2.13%	6.93%	18.35%	26.17%
g'	2.77%	5.62%	9.51%	1.58%
e	0.00%	−0.01%	−0.19%	−0.17%
e_r	2×10^{-4}	2×10^{-3}	2×10^{-2}	3×10^{-2}

Although the approximations by $g' = \mu - \sigma^2/2$ are mostly reasonable, the results do suggest that the errors are larger for assets with higher return volatilities: stocks and commodities in the present case. This makes sense, since, as we argued previously, the error might be in the order of $O(\mu\sigma^2)$. For stocks, both μ and σ are higher. In addition, when g is of similar magnitude to μ, the relative error will be in the order of $O(\sigma^2)$.

Thus, the study of high-volatility assets warrants additional attention. Instead of new asset classes, we shall use leveraged portfolios of the existing assets for this purpose. Former discussion of leveraged portfolios will be the topic of the later chapters. For the time being, we will just use leveraged investments in commodities to study the approximation of the geometric mean. The leverage is achieved by borrowing (or shorting) additional cash and investing the proceeds in the GSCI index. The return of the leveraged portfolio would be

$$r_l = Lr - (L-1)r_f$$

The parameter L stands for leverage, r is the return of the risky asset, and r_f is the return of the risk-free asset.

When $L = 1$, there is no borrowing. If $L = 2$, i.e. a leverage ratio of 2:1, the amount of borrowing is equal to the amount of investment capital.

We use GSCI as the risky asset. As the leverage ratio L increases, the return volatility of r_l is expected to increase proportionally. The question is: how would the accuracy of $g' = \mu - \sigma^2/2$ change? Figure 2.1 displays the geometric returns of the leveraged commodity investments as the leverage increases from one to two, together with the approximation g' and other more accurate approximations based on Equation 2.16. We make several remarks.

- First, the geometric return of the leverage investment declines as leverage increases. GSCI without leverage had an annual return close to 8%. When the leverage reaches 200%, the return is down to 1%. This appears to be counter-intuitive, since the commodity return of 8% is higher than the return of cash, which is used as financing. The explanation for this negative leverage effect is, indeed, the subject of this book. In short, the leverage investment is rebalanced annually, which generates negative rebalancing alpha. However, this is ahead of our discussion. Now, we just use the return of the leveraged investments to test the accuracy of various approximations of the geometric returns.

- When the leverage is low, the approximation by g' is quite good. However, as the leverage increases, the accuracy of the approximation declines steadily. When the leverage reaches two, g is close to 1%, while g' is near 5%. Also, note that g' is higher than g instead of being lower than g. This error increases as the volatility of the returns increases with leverage. This could be caused by other terms that have been omitted from g' and carry negative signs, such as negative skewness.

The approximation can be improved if we retain more terms in Equation 2.16. In Figure 2.1, the line marked $g4$ is the approximation using terms up to $k = 4$, which would include skewness and kurtosis in addition to variance. Similarly, the lines marked $g8$ and $g16$ are the approximations using terms up to $k = 8$ and $k = 16$, respectively. As the number of terms is increased, the approximation improves. It is apparent that $g16$ gives the best result. However, these improved approximations with more terms are too complicated for further analysis of portfolio rebalancing. Our primary objective is to show the limitation of g' as an approximation. Additionally, Figure 2.1 also demonstrates* the validity of our analysis stemming from the expansion in Equation 2.16.

---

* It can be observed in the graph that the approximation of $g16$ deteriorates somewhat when the leverage is close to 1.9. This is likely due to the fact that one of returns violates the constraint in Equation 2.14 at high leverage. Hence, the Taylor expansion (Equation 2.13) is invalid for this return, and the chain of process leading to Equation 2.16 is broken. In fact, the series in Equation 2.16 might not converge at all.

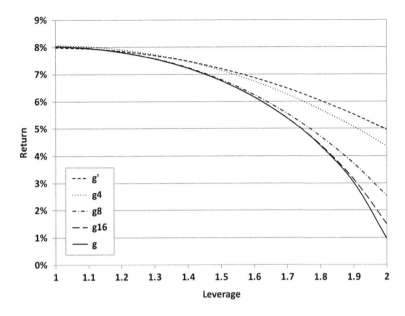

FIGURE 2.1 Geometric return and approximate returns as the leverage increases.

2.4 PORTFOLIO RETURN AND VOLATILITY

So far, our discussion of arithmetic and geometric returns, return volatility, and the relationship between them has been concerned with single or individual assets. We now introduce notations for portfolios of assets in terms of portfolio weights, portfolio returns, and portfolio volatilities. These materials are drawn from standard portfolio theory for portfolios with fixed weights.

Suppose we have M investable assets, and we define a portfolio of assets with a set of portfolio weights $(w_1, w_2, \cdots, w_M)$, which can be denoted by a weight vector (in column) $\mathbf{w} = (w_1, \cdots, w_M)'$. The portfolio weights sum to 1:

$$w_1 + w_1 + \cdots + w_M = 1 \qquad (2.19)$$

The budget constraint (Equation 2.19) can be written in a vector form, $\mathbf{w}' \cdot \mathbf{i} = 1$, with the definition of the vector $\mathbf{i} = (1,1,\ldots,1)'$. The portfolio weights are real numbers; they are non-negative for long-only portfolios, and some of them are negative for long-short portfolios. The formulation and results presented here should apply to both long-only and long-short portfolios.

In period i, the returns of these assets are $(r_{1i}, r_{2i}, \ldots, r_{Mi})$, which can also be denoted by a return vector $\mathbf{r}_i = (r_{1i}, r_{2i}, \ldots, r_{Mi})'$. Then, the portfolio return in period i is

$$R_i = w_1 r_{1i} + w_2 r_{2i} + \cdots + w_M r_{Mi} = \sum_{j=1}^{M} w_j r_{ji} = \mathbf{w}' \cdot \mathbf{r}_i \qquad (2.20)$$

The last expression is the inner product of two vectors.

The arithmetic mean of portfolio returns over N periods is

$$\mu = \frac{R_1 + \cdots + R_N}{N} = w_1 \frac{r_{11} + \cdots + r_{1N}}{N} + \cdots + w_M \frac{r_{M1} + \cdots + r_{MN}}{N} \qquad (2.21)$$

$$= w_1 \mu_1 + \cdots + w_M \mu_M$$

In other words, the arithmetic mean of a portfolio is the weighted average of arithmetic means of individual assets. If we denote the individual means by a vector $\boldsymbol{\mu} = (\mu_1, \mu_2, \ldots, \mu_M)'$, then Equation 2.21 can be written as $\mu = \mathbf{w}' \cdot \boldsymbol{\mu}$. We shall leave the definition of the geometric mean of a portfolio to the next chapter.

The portfolio volatility or variance is defined similarly, and we have

$$\sigma^2 = \frac{1}{N} \left[(R_1 - \mu)^2 + (R_2 - \mu)^2 + \cdots + (R_N - \mu)^2 \right] \qquad (2.22)$$

Using the vector forms for the portfolio returns and the portfolio arithmetic mean in the equation leads to

$$\sigma^2 = \frac{1}{N} \left[(\mathbf{w}' \cdot \mathbf{r}_1 - \mathbf{w}' \cdot \boldsymbol{\mu})^2 + \cdots + (\mathbf{w}' \cdot \mathbf{r}_N - \mathbf{w}' \cdot \boldsymbol{\mu})^2 \right]$$

$$= \frac{1}{N} \left[\mathbf{w}'(\mathbf{r}_1 - \boldsymbol{\mu})(\mathbf{r}_1 - \boldsymbol{\mu})' \mathbf{w} + \cdots + \mathbf{w}'(\mathbf{r}_N - \boldsymbol{\mu})(\mathbf{r}_N - \boldsymbol{\mu})' \mathbf{w} \right] \qquad (2.23)$$

$$= \mathbf{w}' \Sigma \mathbf{w}$$

The matrix Σ is the sample covariance matrix of asset returns:

$$\Sigma = \frac{1}{N} \left[(\mathbf{r}_1 - \boldsymbol{\mu})(\mathbf{r}_1 - \boldsymbol{\mu})' + \cdots + (\mathbf{r}_N - \boldsymbol{\mu})(\mathbf{r}_N - \boldsymbol{\mu})' \right] \qquad (2.24)$$

The $M \times M$ matrix can be expressed in terms of their elements:

$$\Sigma = \begin{pmatrix} \sigma_1^2 & \sigma_{12} & \cdots & \sigma_{1M} \\ \sigma_{21} & \sigma_2^2 & \cdots & \sigma_{2M} \\ \vdots & \vdots & \ddots & \vdots \\ \sigma_{M1} & \sigma_{M2} & \cdots & \sigma_M^2 \end{pmatrix} \tag{2.25}$$

The covariance matrix is symmetric and positive definite unless $\sigma = 0$. The diagonal elements are variances of individual assets, which are defined as Equation 2.6 and its variants in Equations 2.7 and 2.8. The off-diagonal elements are covariances between pairwise assets, given by

$$\sigma_{ij} = \frac{1}{N} \sum_{k=1}^{N} \left(r_{ik} - \mu_i \right) \left(r_{jk} - \mu_j \right) \tag{2.26}$$

It is customary to define covariance by the symbol cov, such that

$$\sigma_{ij} = \text{cov}\left(\vec{r_i}, \vec{r_j} \right) \tag{2.27}$$

where the vectors $\vec{r_i}, \vec{r_j}$ are returns for assets i and j over N periods. Then, the variance of individual assets is $\sigma_i^2 = \text{var}\left(\vec{r_i} \right) = \text{cov}\left(\vec{r_i}, \vec{r_i} \right)$

The portfolio variance can be rewritten in terms of variances and covariances, and we have

$$\sigma^2 = \mathbf{w}'\Sigma\mathbf{w} = \sum_{i,j=1}^{M} w_i w_j \sigma_{ij} = \sum_{i=1}^{M} w_i^2 \sigma_i^2 + 2 \sum_{i<j} w_i w_j \sigma_{ij} \tag{2.28}$$

The first summation consists of variances from individual assets, and the second summation consists of all covariances of pairs formed with different assets.

The covariance matrix can also be written in terms of volatilities and correlation matrix. First, we define pairwise correlation:

$$\rho_{ij} = \frac{\sigma_{ij}}{\sigma_i \sigma_j} \tag{2.29}$$

It is easy to show that $|\rho_{ij}| \leq 1$. Then,

$$\Sigma = \text{diag}(\sigma_1,\ldots,\sigma_M) \, C \, \text{diag}(\sigma_1,\ldots,\sigma_M) \tag{2.30}$$

The matrix C is the correlation matrix $C = (\rho_{ij})_{i,j=1}^{M}$, and $\text{diag}(\sigma_1,\ldots,\sigma_M)$ is a diagonal matrix with volatilities of individual assets as its diagonal elements. Using correlations, Equation 2.28 can be written as

$$\sigma^2 = \mathbf{w} = \sum_{i=1}^{M} w_i^2 \sigma_i^2 + 2 \sum_{i<j} w_i w_j \sigma_i \sigma_j \rho_{ij} \tag{2.31}$$

Example 2.4: Consider the case of two assets. The portfolio variance becomes

$$\sigma^2 = w_1^2 \sigma_1^2 + 2\rho_{12} w_1 w_2 \sigma_1 \sigma_2 + w_2^2 \sigma_2^2 \tag{2.32}$$

When both weights are non-negative, we have the following special cases of correlations:

$$\sigma = \begin{cases} w_1 \sigma_1 + w_2 \sigma_2 & \text{if } \rho_{12} = 1 \\ \sqrt{w_1^2 \sigma_1^2 + w_2^2 \sigma_2^2} & \text{if } \rho_{12} = 0 \\ |w_1 \sigma_1 - w_2 \sigma_2| & \text{if } \rho_{12} = -1 \end{cases} \tag{2.33}$$

The portfolio volatility is highest when the correlation is 1, and it is lowest when the correlation is negative. The lower the correlation, the lower the volatility. The diversification benefit increases with declining correlation.

Example 2.5: We can also use Equation 2.32 to derive the volatility of return differences between two assets by setting $w_1 = 1$ and $w_2 = -1$. Then,

$$\sigma_{1-2}^2 = \sigma_1^2 - 2\rho_{12}\sigma_1\sigma_2 + \sigma_2^2 \tag{2.34}$$

Note that we have used σ_{1-2} to denote the volatility of return differences.

It might appear to some readers that the weights of Example 2.5 are not right for an actual portfolio, because they do not sum to 1. This is true in a strict sense. However, we can make this an actual long-short portfolio by investing 100% (long) in Asset 1, −100% (short) in Asset 2, and 100% (long) in risk-free assets. Then, the sum of the weights is 1, and the volatility of this long-short portfolio will be given by Equation 2.34.

Example 2.6: Another special kind of long-short portfolio is a levered long-only portfolio. We start with a long-only portfolio with positive weights $\mathbf{w} = (w_1, \ldots, w_M)'$ and $w_1 + w_1 + \cdots + w_M = 1$. Denote Asset 0 as the risk-free asset. A levered portfolio can be constructed by borrowing additional funds from the risk-free asset and investing them in the long-only portfolio. Suppose $w_0 < 0$ is the weight of the risk-free assets. The levered portfolio would have weights

$$\mathbf{w}^L = \left[w_0, \left(1 - w_0\right) w_1, \ldots, \left(1 - w_0\right) w_M \right]' \tag{2.35}$$

Since the risk-free asset has no return volatility, the volatility of the levered portfolio is $\left(1 - w_0\right)$ times the volatility of the original long-only portfolio. The quantity $L = \left(1 - w_0\right)$ is the leverage of the portfolio. We point out that the result is equally valid if $w_0 > 0$. When $0 < w_0 < 1$, the final portfolio weights (Equation 2.35) are all positive. When $w_0 > 1$, the final portfolio is long on risk-free assets and short on the original portfolio.

2.5 SERIAL CORRELATION AND VOLATILITY OF MULTI-PERIOD RETURNS

The analysis of asset returns and volatilities depends on the investment horizon. In this section, we are concerned with the volatility of multi-period returns for a single asset and a portfolio of assets.

For a single asset whose returns are serially independent, the volatility of multi-period returns would be approximately the volatility of a single-period return multiplied by the square root of the number of periods. For example, consider the returns of two periods. When the correlation between two subsequent periods is zero, the volatility of a two-period return would be $\sqrt{2}\sigma$, where σ is the volatility of single-period returns. This result follows the second case of Equation 2.33 once we take $w_1 = w_2 = 1$ and assume $\sigma_1 = \sigma_2 = \sigma$.

TABLE 2.4   Serial Autocorrelations of Annual Asset Returns
from 1970 to 2014

	Cash	Bonds	Stocks	Commodities
$\rho(1)$	0.89	−0.04	0.00	0.09
$\rho(2)$	0.71	−0.09	−0.19	−0.16
$\rho(3)$	0.57	0.10	0.09	−0.08

However, in reality, asset returns are often serially correlated. Their behavior often exhibits momentum or reversal, and momentum is associated with positive serial correlation, while reversal is associated with negative serial correlation. Table 2.4 lists the serial autocorrelations of four asset classes in the previous discussions. They are of different lags, from one to three years. $\rho(1)$ is the correlation between return in one year and return in the subsequent year; $\rho(2)$ is the correlation between return in one year and return in the year after the subsequent year; and so on.

The pattern is different for the different assets. The correlation is all positive for cash, whose return is tied to short-term interest, which is very persistent. The first two correlations for bonds are negative, implying return reversal. This is partly due to the reversal of bond yields. The correlation with lag of three years is positive. For stocks, the one-year serial correlation is zero. However, it is significantly negative for the two-year lag and turns positive for the three-year lag. Commodities are the only risky asset with one-year momentum, since its one-year serial correlation is positive. However, it reverses in years two and three.

2.5.1 Single Asset Multi-Period Volatility

For a single asset, when the correlation is not zero, we obtain, using Equation 2.32,

$$\sigma\langle 2 \rangle = \sigma\sqrt{2 + 2\rho(1)} \tag{2.36}$$

We have used $\rho(1)$ to denote the serial correlation of lag 1 and $\sigma\langle 2 \rangle$ to denote the volatility of two-period returns. Note that we have implicitly assumed that the two-period return is the sum of two one-period returns. In practice, the two-period return is the cumulative return of two one-period returns. Hence, Equation 2.36 is at best an approximation.

In general, we are interested in the volatility of multi-period return:

$$\sigma\langle H\rangle = \sigma\left(r_1 + \cdots + r_H\right) \tag{2.37}$$

This volatility is similar in form to the volatility of a portfolio in which all assets have the same volatility, and all portfolio weights are 1. Hence, by Equation 2.31, we have

$$\sigma^2\langle H\rangle = \sigma^2\left[H + 2\sum_{h<g}\rho(h,g)\right] \tag{2.38}$$

The correlation $\rho(h,g)$ denotes the serial correlation between r_g and r_h. Assuming the returns are stationary, the serial correlation only depends on the lag between the two periods, and $\rho(h,g) = \rho(h-g) = \rho(g-h)$. Equation 2.38 can be rewritten as

$$\sigma\langle H\rangle = \sigma\sqrt{H + 2\sum_{h=1}^{H-1}(H-h)\rho(h)} \tag{2.39}$$

The coefficient of $\rho(h)$ reflects the number of return pairs with lag h. When the lag is small, there are more pairs; when the lag is large, there are fewer pairs. The variance of multi-period returns also has a matrix form. The derivation is left as an exercise.

Equation 2.39 also leads to volatility of average return over multiple periods.

We have

$$\sigma(\mu) = \sigma\left(\frac{r_1 + \cdots + r_H}{H}\right) = \frac{\sigma}{\sqrt{H}}\sqrt{1 + 2\sum_{h=1}^{H-1}\left(1 - \frac{h}{H}\right)\rho(h)} \tag{2.40}$$

The volatility of the arithmetic mean in general declines as the horizon lengthens, inversely proportional to the square root of the number of periods. Serial correlation influences it similarly to the case of multi-period return variance.

For theoretical time series analysis, $\rho(h)$ might have analytic expressions based on particular time series models. We have an exercise for the $AR(1)$ process. In empirical analysis, we can take $\rho(h)$ as the average of serial correlations with the same lag h.

2.5.2 Portfolio Multi-Period Volatility

The multi-period volatility of a single asset is affected by its serial auto-correlations. For a portfolio, however, both serial autocorrelation of individual assets and serial cross-correlations between different assets have an impact on its multi-period return volatility. Let us first consider a portfolio of two assets and a horizon of two periods. Using the notation of Equation 2.20, we have

$$R_1 = w_1 r_{11} + w_2 r_{21}, R_2 = w_1 r_{12} + w_2 r_{22}$$

Then,

$$\sigma^2 \langle 2 \rangle = \text{var}\left(w_1 r_{11} + w_2 r_{21} + w_1 r_{12} + w_2 r_{22} \right) \tag{2.41}$$

The variance on the right-hand side of Equation 2.41) is the variance of a sum of four return terms. The calculation can proceed with its expansion into 16 covariance terms. Covariances such as $\text{cov}(r_{11}, r_{12})$ and $\text{cov}(r_{21}, r_{22})$ will contain serial autocorrelations of the two assets. Covariances such as $\text{cov}(r_{11}, r_{22})$ and $\text{cov}(r_{21}, r_{12})$, on the other hand, will involve serial cross-correlations between the two assets. For example, $\text{cov}(r_{11}, r_{22})$ is the covariance between the return of the first asset in Period 1 and the return of the second asset in Period 2.

The expression resulting from Equation 2.41 is rather cumbersome. We shall present a derivation based on return vectors and covariance matrices. As previously in the chapter, let $\mathbf{r}_1 = \left(r_{11}, \ldots, r_{M1} \right)'$ be the return vector in period 1, $\mathbf{r}_2 = \left(r_{12}, \ldots, r_{M2} \right)'$ being the return vector in period 2, and so on. Let $\mathbf{w} = \left(w_1, \ldots, w_M \right)'$ be the portfolio weight. Then, the variance of multi-period return is

$$\sigma^2 \langle H \rangle = \text{var}\left[\mathbf{w}' \cdot \left(\mathbf{r}_1 + \cdots + \mathbf{r}_H \right) \right] \tag{2.42}$$

We then have

$$\text{var}\left[\mathbf{w}' \cdot \left(\mathbf{r}_1 + \cdots + \mathbf{r}_H \right) \right] = \mathbf{w}' \Sigma \langle H \rangle \mathbf{w} \tag{2.43}$$

where $\Sigma \langle H \rangle$ is the covariance matrix of $\mathbf{r}_1 + \cdots + \mathbf{r}_H$. The covariance matrix $\Sigma \langle H \rangle$ equals a sum of covariance matrices:

$$\Sigma\langle H\rangle = \sum_{h,g=1}^{H}\Sigma_{h,g} \tag{2.44}$$

Finally, $\Sigma_{h,g}$ is the covariance matrix between $\mathbf{r}_h$—returns in period h—and $\mathbf{r}_g$—returns in period g. We can write $\Sigma_{h,g} = \mathrm{cov}\left(\mathbf{r}_h,\mathbf{r}_g\right)$. To summarize, the variance of multi-period return is given by

$$\sigma^2\langle H\rangle = \mathbf{w}'\left(\sum_{h,g=1}^{H}\Sigma_{h,g}\right)\mathbf{w} \tag{2.45}$$

When the two subscripts are the same, the two returns vectors are the same, and we have the regular "contemporaneous" covariance matrix, i.e. $\Sigma_{h,g} = \Sigma$. In those cases, we have $\mathbf{w}'\Sigma\mathbf{w} = \sigma^2$—the volatility of single-period portfolio return. Therefore,

$$\sigma^2\langle H\rangle = H\sigma^2 + 2\mathbf{w}'\left(\sum_{h<g}^{H}\Sigma_{h,g}\right)\mathbf{w} \tag{2.46}$$

Note that Equation 2.46 is analogous to Equation 2.38.

PROBLEMS

2.1. Suppose an investment has a cumulative return of 0% after two periods, i.e. $\left(1+r_1\right)\left(1+r_2\right)-1=0$. (a) Show that the two returns are of opposite sign. (b) Show that $r_1 + r_2 \geq 0$.

2.2. (a) Prove the identity in Equation 2.8. (b) Prove the identity in Equation 2.11.

2.3. (a) Prove that a better approximation of the geometric return based on Equation 2.17 is

$$g \approx \mu - \frac{\sigma^2}{2} + \frac{\mu\sigma^2}{2}$$

(b) Improve this approximation further with inclusion of skewness.

2.4. Show that $\left|\rho_{ij}\right| \leq 1$.

2.5. Show that Equation 2.39 can be rewritten as $\sigma^2 \langle H \rangle = \sigma(\mathbf{i}'\mathbf{C}_H \mathbf{i})$, where the vector $\mathbf{i}$ is the vector of ones, and the correlation matrix is given by

$$
\mathbf{C}_H = \begin{pmatrix}
1 & \rho(1) & \cdots & \rho(H-1) \\
\rho(1) & 1 & \cdots & \rho(H-2) \\
\vdots & \vdots & \ddots & \vdots \\
\rho(H-1) & \rho(H-2) & \cdots & 1
\end{pmatrix}
$$

The element $C_{ij} = \rho(|i-j|)$.

2.6. For an AR(1) process, the autocorrelation function is $\rho(h) = \phi^h$. Derive an analytic expression for $\sigma(H)$ based on Equation 2.39.

2.7. Expand the variance in Equation 2.41 explicitly in terms of variance and covariance of asset returns.

2.8. We denote a stacked return vector $\mathbf{R}$ as

$$
\mathbf{R} = \begin{pmatrix}
\mathbf{r}_1 \\
\mathbf{r}_2 \\
\vdots \\
\mathbf{r}_H
\end{pmatrix}
$$

2.9. What is the covariance matrix of $\mathbf{R}$? (b) Write the variance of multi-period portfolio return (Equation 2.46) in terms of this "grand" covariance matrix.

CHAPTER 3

Portfolio Rebalancing

IN THIS CHAPTER, WE introduce the concept and process of portfolio rebalancing and define rebalancing alpha in mathematical terms that will be used in quantitative analyses in later chapters. Along the way, we provide some simple analytical examples to illustrate the process of portfolio rebalancing for both long-only and long-short portfolios. Near the end of the chapter, we provide empirical examples, which could have either positive or negative rebalancing alpha. Understanding and distinguishing these cases is the motivation for the rest of the book.

3.1 SIMPLE EXAMPLES

The following example provides an interesting illustration of the benefit of portfolio rebalancing. Suppose we have two investments at our disposal and an investment horizon of two years. The first investment, labelled Investment A, returns 100% in the first year and −50% in the second year. This is the return pattern in Example 2.1, when we discussed the difference between arithmetic and geometric returns. Investment A has a geometric of zero over the two-year period. The second investment, labelled Investment B, has a return of −50% in the first year and 100% in the second year. Like Investment A, Investment B also has no return to show for two tumultuous years. Suppose we have an initial capital of $1, and we invest 50 cents each in Investments A and B. We now consider the investment results of two different approaches: one is buy-and-hold, and the other is fixed-weight with portfolio rebalancing.

TABLE 3.1   Investment History of a Buy-and-Hold Portfolio

	$t = 0$	$t = 1$	$t = 2$
Investment A	$0.5	$1 (80%)	$0.5
Investment B	$0.5	$0.25 (20%)	$0.5
Portfolio total	$1	$1.25	$1

It is quite simple to see the investment result of buy-and-hold, whereby we hold the initial investment until the end of two years. Since both Investments A and B maintain their value over the whole two-year period, the return of buy-and-hold of this 50/50 portfolio is zero. In fact, the return of any buy-and-hold portfolio is zero. We write $r_{BH} = 0\%$.

We note that the return of this portfolio is not zero after the first year. As shown in Table 3.1, the portfolio value increased from $1 to $1.25 after the first year, of which $1 was in Investment A and $0.25 in Investment B. Only after the second year did both investments revert to $0.5.

Notice that the portfolio weights changed. At the end of the first year, due to return difference, Investment A accounted for 80% of the portfolio, while Investment B accounted for only 20%. For a fixed-weight portfolio, the process requires rebalancing portfolio weights back to the original 50% in both A and B.

To rebalance the portfolio after Year 1 to the original weights, both Investments A and B should be at $0.625. This is explicitly shown in Table 3.2. In the following year, investment A shrank to $0.3125, while investment B grew to $1.25, leaving the total portfolio at a final value of $1.5626. To summarize the results, the buy-and-hold portfolio has a return of zero, and the fixed-weight portfolio has a return of 56.25% (25% per year) over the two-year period.

- This is an exaggerated example, since the two investments are mostly identical, except that they have perfect negative correlation.

- When the two investments have positive correlation, the fixed-weight portfolio could still generate positive return when both investments have zero return over the two-year period (see Problem 2.1).

- Note that portfolio rebalancing after Year 1 is achieved by selling Investment A (the winner in Year 1) and buying Investment B (the loser in Year 1). We shall see later in the chapter that this is the general mechanism of portfolio rebalancing for long-only portfolios.

TABLE 3.2   Investment History of the Fixed-Weight Portfolio

	$t = 0$	$t = 1$ (before)	$t = 1$ (after)	$t = 2$
Investment A	$0.5	$1 ($w_A = 80\%$)	$0.625 ($w_A = 50\%$)	$0.3125
Investment B	$0.5	$0.25 ($w_B = 20\%$)	$0.625 ($w_B = 50\%$)	$1.25
Portfolio total	$1	$1.25	$1.25	$1.5625

TABLE 3.3   Investment History of another Buy-and-Hold Portfolio

	$t = 0$	$t = 1$	$t = 2$
Investment C	$0.5	$0.575 (52%)	$0.6613
Investment D	$0.5	$0.525 (48%)	$0.5513
Portfolio total	$1	$1.1	$1.2125

Our last comment indicates that if the winner in Year 1 keeps on winning, a buy-and-hold portfolio would have a better investment result than a fixed-weight portfolio. To illustrate this point, we consider another two investments, C and D, with C returning 15% each year and D returning 5% each year. Table 3.3 displays the investment results of a 50/50 buy-and-hold portfolio with these two investments. Note that after Year 1, the portfolio weights have drifted to 52% in C and 48% in D. Without rebalancing, this portfolio earned a compound return of 21.25% over the two-year period.

What happens if we rebalance the portfolio after Year 1? Table 3.4 shows the results. The final result for the fixed-weight portfolio is a compound return of 21% over the two-year period, which is 0.25% or 25 basis points less than the return of the buy-and-hold portfolio. In hindsight, the underperformance is due to the fact that Investment C outperformed Investment D in Year 2, and the fixed-weight portfolio had less weight in Investment C than the buy-and-hold portfolio did. It is easy to calculate the performance difference as ($2.27\% \cdot 10\% \cdot 1.1 = 0.25\%$).

TABLE 3.4   Investment History of another Fixed-Weight Portfolio

	$t = 0$	$t = 1$ (before)	$t = 1$ (after)	$t = 2$
Investment C	$0.5	$0.575 ($w_C = 52\%$)	$0.55 ($w_C = 50\%$)	$0.6325
Investment D	$0.5	$0.525 ($w_D = 48\%$)	$0.55 ($w_D = 50\%$)	$0.5775
Portfolio total	$1	$1.1	$1.1	$1.2100

These two simple and extreme examples provide insights into the effects of portfolio rebalancing. However, in reality, no asset returns behave like the investments in the examples. In most cases, their characteristics are a combination of positive returns over time (C and D), intra-period return volatilities (A and B), and different kinds of serial correlations. One of the main objectives of the present book is to determine the relative return of buy-and-hold and fixed-weight portfolios in terms of realistic return characteristics.

3.2 REBALANCING LONG-ONLY PORTFOLIOS

The previous examples show that rebalancing long-only portfolios leads to buying low return assets and selling high return assets. We now formally generalize these observations. We emphasize that mathematically, there is little difference between the formulation of long-only and long-short portfolios. However, as we shall see in the next section, the intuition and mechanism of rebalancing long-short portfolios are quite different from those for long-only portfolios.

We consider a portfolio of M assets with weights, $\mathbf{w} = \left(w_1, \cdots, w_M\right)'$ and $\sum_{i=1}^{M} w_i = 1$. Suppose returns of the assets in a period are $\mathbf{r} = \left(r_1, r_2, \ldots, r_M\right)'$. The portfolio return is $R = w_1 r_1 + w_2 r_2 + \cdots + w_M r_M = \mathbf{w}' \cdot \mathbf{r}$. We have dropped the subscript i for the time being, since we are only studying the drift of portfolio weights after a set of returns.

The portfolio weights have drifted to

$$w_i^d = \frac{w_i\left(1 + r_i\right)}{1 + R}, i = 1, \ldots, M \tag{3.1}$$

Note that $\sum_{i=1}^{M} w_i^d = 1$. And the change of portfolio weights is

$$\Delta w_i = w_i^d - w_i = \frac{w_i\left(r_i - R\right)}{1 + R}, i = 1, \ldots, M \tag{3.2}$$

For long-only portfolios, we have

$$\begin{cases} \Delta w_i \geq 0, \text{ if } r_i \geq R \\ \Delta w_i \leq 0, \text{ if } r_i \leq R \end{cases} \tag{3.3}$$

In other words, assets that outperformed the portfolio would have increases in weight, and assets that underperformed the portfolio would have decreases in weight. In addition, the change in the weight of an asset is proportional to both the relative performance of the asset against the portfolio and the original weight.

Example 3.1: Consider a portfolio with two assets. According to Equation 3.2, we have

$$\Delta w_1 = \frac{w_1\left(r_1 - R\right)}{1+R} = \frac{w_1 w_2\left(r_1 - r_2\right)}{1 + w_1 r_1 + w_2 r_2}, \Delta w_2 = -\Delta w_1 \tag{3.4}$$

When both weights are positive, the asset with higher return will have an increase in its portfolio weight. It is also obvious from Equation 3.4 that there is no weight change if one of the weights is zero or the two returns are the same.

For a given set of returns, the changes of portfolio weights vary with the initial portfolio weights. It is important to study the magnitude of weight changes, because they are related to the turnover of portfolio rebalancing, which will be defined shortly. They are also related to the size of rebalancing alpha. For long-only portfolios in which both weights are non-negative, from Equation 3.4, we note that the numerator is maximized when $w_1 = w_2 = 50\%$. However, since the denominator also depends on the weights, the maximum of Equation 3.4 will not be exactly achieved by the equal-weight portfolio, although the deviation is usually quite small (see Problem 3.2).

We now define portfolio turnover caused by rebalancing, which is the sum of absolute values of weight changes:

$$T = \frac{1}{2}\sum_{i=1}^{M}\left|\Delta w_i\right| \tag{3.5}$$

This definition measures the one-way (either buys or sells) turnover of the portfolio rebalancing. Substituting Equations 3.2 into Equation 3.5 leads to

$$T = \frac{1}{2}\sum_{i=1}^{M}\frac{\left|w_i\right|\left|r_i - R\right|}{1+R} \approx \frac{1}{2}\sum_{i=1}^{M}\left|w_i\right|\left|r_i - R\right| \tag{3.6}$$

The approximation is valid where the portfolio return is low.

For a fixed-weight portfolio, the turnover changes from period to period, depending on the realized returns. The expected turnover or average turnover over time can be approximated by the expectation

$$E(T) \approx \frac{1}{2} \sum_{i=1}^{M} |w_i| E(|r_i - R|) \tag{3.7}$$

To make further progress, we assume that asset returns are normally distributed. Consequently, the return difference between the asset i and the portfolio follows a normal distribution:

$$r_i - R \sim N\left(\mu_i - \mu, \sigma_{i-R}^2\right) \tag{3.8}$$

We have assumed that μ_i and μ are average returns of the asset i and the portfolio, respectively. The variance of return difference (Example 2.5) is

$$\sigma_{i-R}^2 = \sigma_i^2 + \sigma^2 - 2\rho_{i,p}\sigma_i\sigma \tag{3.9}$$

Note that σ_i and σ are the volatilities of the asset i and the portfolio, respectively, and $\rho_{i,p}$ is the correlation between the two. When the average returns are equal, or when their difference is small compared with the volatility of their return difference, the expectation of the absolute return difference can be derived analytically. We use the result as an approximation of the average turnover. We have

$$E(T) \approx \frac{1}{\sqrt{2\pi}} \sum_{i=1}^{M} |w_i| \sigma_{i-R} \tag{3.10}$$

It is easy to see that the volatility σ_{i-R} is a declining function of the correlation $\rho_{i,p}$ between individual assets and the portfolio. When all individual assets are highly correlated with each other, they tend to be highly correlated with the portfolio. This would lead to lower turnover.

Example 3.2: For a portfolio with two assets,

$$\sigma_{1-R} = |w_2|\sigma_{1-2}, \text{ and } \sigma_{2-R} = |w_1|\sigma_{1-2} \tag{3.11}$$

Hence,

$$E(T) \approx \sqrt{\frac{2}{\pi}} |w_1 w_2| \sigma_{1-2} = \sqrt{\frac{2}{\pi}} |w_1 w_2| \sqrt{\sigma_1^2 + \sigma_2^2 - 2\rho_{12}\sigma_1\sigma_2} \qquad (3.12)$$

The advantage of Equation 3.12 is that it expresses turnover in terms of volatilities of individual assets and the correlation between them rather than those involving the portfolio.

Example 3.3: We use Equation 3.12 to estimate the expected turnover of portfolio rebalancing of two-asset allocation portfolios, comprised of the four assets discussed in Chapter 2. In Table 3.5, we list their volatilities and correlation matrix based on annual returns from 1970 to 2014. Let us first consider a portfolio of two assets, one of which is cash. Note that cash return volatility is quite low compared with other asset classes. As a result, it contributes little to the volatility of return difference. Hence, we can use the volatility of risky asset for the volatility of return difference, i.e. $E(T) \approx \sqrt{2/\pi} |w_1 w_2| \sigma_2$ where we assume that Asset 1 is the cash and Asset 2 is a risky asset. By this formula, the expected turnover of a 50/50 cash/bond portfolio would be

$$E(T) \approx \sqrt{2/\pi} \cdot 0.5 \cdot 0.5 \cdot 7.6\% = 1.52\%$$

According to similar calculations, the expected turnover of a 50/50 cash/stock portfolio is 3.44%, and that of a 50/50 cash/commodity portfolio is 4.92%.

For a stock/bond portfolio, the volatility of their return difference is

$$\sigma_{1-2} = \sqrt{(7.6\%)^2 + (17.24\%)^2 - 2(0.02)(7.6\%)(17.24\%)} = 18.72\%$$

TABLE 3.5 Volatilities and Correlation Matrix of the Four Asset Classes

	Cash	Bonds	Stocks	Commodities
σ	3.38%	7.60%	17.24%	24.66%
Cash	1.00	0.30	0.04	0.21
Bonds	0.30	1.00	0.02	−0.17
Stocks	0.04	0.02	1.00	−0.08
Commodities	0.21	−0.17	−0.08	1.00

This is slightly higher than the volatility of stock return alone. Substituting it into Equation 3.12, we obtain the turnover of a 50/50 stock/bond portfolio as 3.73%.

3.3 REBALANCING LONG-SHORT PORTFOLIOS

The analysis of portfolio rebalancing, up to this point, is valid for both long-only and long-short portfolios. For example, in discussion of portfolio turnover, as long as we take the absolute value of portfolio weights, the results from Equations 3.6 through 3.12 are true for long-short portfolios. However, there is one major difference between portfolio rebalancing of long-only and long-short portfolios. The difference lies in the mechanism of the rebalancing process. In long-only portfolios, rebalancing amounts to buying losers and selling winners. But in long-short portfolios, this is not always the case. This difference is crucial when we study the effects of portfolio rebalancing and rebalancing alpha for long-short portfolios. In this section, we want to gain some insights into the rebalancing process of long-short portfolios.

The best place to start is to consider a two-asset long-short portfolio with $w_1 < 0$, $w_2 > 0$, and $w_1 + w_2 = 1$. The leverage of the portfolio is the sum of the absolute value of portfolio weights:

$$L = |w_1| + |w_2| = w_2 - w_1 > 1 \tag{3.13}$$

Then, by Equation 3.4,

$$\Delta w_1 = \frac{w_1 w_2 \left(r_1 - r_2 \right)}{1+R}, \Delta w_2 = \frac{w_1 w_2 \left(r_2 - r_1 \right)}{1+R} \tag{3.14}$$

We assume that the portfolio is not bankrupt, i.e. $1 + R > 0$. Note that since $w_1 w_2 < 0$, the change in portfolio weight of each asset is now negatively proportional to the relative performance of the particular asset versus the other asset. In other words, if an asset outperforms the other asset, then its weight change is negative, or its weight will be lower than the initial weight. To rebalance the portfolio, we need to buy the asset. On the other hand, if an asset underperforms the other asset, then its weight change is positive, or its weight will be higher than the initial weight. To rebalance the portfolio, we need to sell the asset. In summary, to rebalance this long-short portfolio, we buy the winner and sell the loser, which is the opposite action to rebalancing a long-only portfolio.

TABLE 3.6   Investment History of another Buy-and-Hold Portfolio

	$t=0$	$t=1$ (before)	$t=1$ (after)
Investment A	$1.2 ($w_A = 120\%$)	$2.4 ($w_A = 104\%$)	$2.74(120\%)
Investment B	$-0.2 ($w_B = -20\%$)	$-0.1 ($w_B = -4\%$)	$-0.46 (-20\%)
Portfolio total	$1	$2.3	$2.3
Portfolio leverage	140%	109%	140%

Let us use a concrete example to illustrate this point. Consider Investments A and B from earlier, with A returning 100% in Year 1 and B returning −50% in Year 1. We construct a long-short portfolio with 120% long Investment A and 20% short in Investment B. Table 3.6 displays the investment results of this portfolio after Year 1. First, note that after Year 1, Investment A grew from $1.2 to $2.4, and Investment B went from −$0.2 to −$0.1. As a result of leverage and correct positioning in both investments, the total investment increased from $1 to $2.3. The portfolio weights are now 104% and −4%, respectively. Compared with the original weights of 120% and −20%, weight in Investment A has decreased, and weight in Investment B has increased. To rebalance the portfolio back to the original weights, we need to increase investment in A to $2.76 (buy winner) and decrease investment in B to −$0.46 (sell loser).

In the last row of Table 3.6, we show the portfolio leverage over time. Before the rebalance, the leverage has declined from the original 140% to 109%. Rebalancing also re-levers the portfolio to 140%.

Things could well turn out differently. In Table 3.7, we display investment results of a portfolio that is long 120% in Investment B and short 20% in Investment A. After one year, both positions are detrimental, with investment in A declining from −$0.2 to −$0.4 and investment in B shrinking from $1.2 to $0.6. The portfolio value becomes $0.2, a loss of 80%. What happened to portfolio weights is much more dramatic. They are now −200% and 300%, respectively, in A and B. To rebalance the portfolio, one needs to

TABLE 3.7   Investment History of another Buy-and-Hold Portfolio

	$t=0$	$t=1$ (Before)	$t=1$ (After)
Investment A	$-0.2 ($w_A = -20\%$)	$-0.4 ($w_A = -200\%$)	$-0.04 (-20\%)
Investment B	$1.2 ($w_B = 120\%$)	$0.6 ($w_B = 300\%$)	$0.24 (120\%)
Portfolio total	$1	$0.2	$0.2
Portfolio leverage	140%	500%	140%

sell Investment B (sell loser) down to $0.24 and buy (cover) Investment A (buy winner) from −$0.4 to −$0.04.

In both cases, portfolio rebalancing entails buying winners and selling losers. We also notice that when the long-short portfolio makes a profit, portfolio leverage declines, and when it suffers a loss, portfolio leverage rises.

For a general portfolio with more than two assets, we use Equation 3.2 to determine the changes in portfolio weights, with a distinction between long positions and short positions. For long positions, the results are similar to Equation 3.3. The weight of a long position increases (decreases) if the asset outperforms (underperforms) the portfolio. For short positions, the results are exactly opposite; we have

$$
\begin{cases}
\Delta w_i \geq 0, \text{ if } r_i \leq R \\
\Delta w_i \leq 0, \text{ if } r_i \geq R
\end{cases}
\tag{3.15}
$$

In other words, the weight of a short position increases (decreases) if the asset underperforms (outperforms) the portfolio. But one must note that an increase in a short position means the weight is becoming less negative. To rebalance the portfolio, one must sell the asset. On the other hand, a decrease in a short position means the weight is getting more negative. One must buy back the asset to rebalance.

Perceptive readers might find that the results of Equation 3.3 for long positions are not entirely consistent with our examples listed in Tables 3.6 and 3.7, where we bought winner in long position and sold loser in long position in the rebalancing process. But there is no contradiction. In both Equations 3.3 and 3.15, the relative performance is against the portfolio. In the examples listed in Tables 3.6 and 3.7, we identify winners and losers rather informally by relative performance between the two assets. Because a long-short portfolio is a levered portfolio, it is possible that all assets underperform (Table 3.6) the portfolio or all assets outperform (Table 3.7) the portfolio. Therefore, a winner versus the other asset could still lose out to the portfolio, and a loser versus the other asset could be a winner against the portfolio. It must be noted that this situation is not possible in a long-only portfolio.

Example 3.4: To illustrate the dynamics of long-short portfolio rebalancing, we provide a four-asset portfolio with three long positions and one short position. This portfolio resembles a levered long-only portfolio in which Asset 1 is the risk-free asset, and the other three are risky assets. The weight vector is displayed in Table 3.8. The leverage of the portfolio is 300%. Suppose the asset returns are 0%, 1%, 6%, and 2%, respectively. The portfolio return is 5%. Notice that the drifted portfolio weights have changed. For Asset 1, a short position, its weight has increased, since its return is less than the portfolio return. For Assets 2 and 4, their returns are also less than the portfolio, and therefore, their weights have declined. However, for Asset 3, its return is higher than the portfolio return, and as a result, its weight has increased. To rebalance the portfolio, we sell Assets 1 and 3 and buy Assets 2 and 4. This rebalance also re-levered the portfolio back to 300% from 290.5%.

A related topic is leverage of a long-short portfolio. As we have mentioned before, the leverage is the sum of absolute values of portfolio weights, which is always 1 for long-only portfolios and greater than 1 for long-short portfolios. Therefore, a long-only portfolio would have constant leverage, and a long-short portfolio's leverage can change. The change in leverage, if left without portfolio rebalancing, could be quite significant over time and have a large impact on a portfolio's risk/return characteristics.

The examples of long-short portfolios we have presented so far seem to indicate that the leverage of a long-short portfolio would decrease or increase if the portfolio had a profit or loss. As we shall now demonstrate, this is not always the case.

For the two-asset long-short portfolio, the leverage is given by $L = w_2 - w_1 > 1$. With drift of portfolio weights, the new leverage is given by

$$L^d = w_2^d - w_1^d = \frac{w_2(1+r_2)}{1+R} - \frac{w_1(1+r_1)}{1+R} = \frac{w_2 - w_1 + w_2 r_2 - w_1 r_1}{1 + w_1 r_1 + w_2 r_2} \quad (3.16)$$

TABLE 3.8   Portfolio Weights and Returns of a Four-Asset Portfolio

	w	r	w^d	Δw
Asset 1	−100%	0.0%	−95.2%	4.8%
Asset 2	20%	1.0%	19.2%	−0.8%
Asset 3	30%	6.0%	30.3%	0.3%
Asset 4	150%	2.0%	145.7%	−4.3%
Portfolio	100%	5.00%	100%	0%
Leverage	300%		290.5%	

The change in leverage is then $\Delta L = L^d - L$. With Equation 3.16, simple algebraic calculation leads to

$$\Delta L = \frac{2w_1 w_2 \left(r_2 - r_1\right)}{1 + w_1 r_1 + w_2 r_2} \tag{3.17}$$

Since $w_1 w_2 < 0$, the sign of change in ΔL is the opposite of the return difference between Asset 2 and Asset 1. When the long outperforms the short, the leverage decreases, and when the long underperforms the short, the leverage increases. In both cases, the portfolio return could be either positive or negative.

The analysis of a general long-short portfolio with more than two assets is much more complicated if we consider changes of all individual asset weights in terms of individual asset returns. One way to simplify the problem is to divide the assets into two groups: one is the group of short assets and the other the group of long assets. Then, the portfolio is short the first group and long the second group, resulting in a two-group long-short portfolio. The analysis of a two-asset portfolio can be used to derive similar results.

We define aggregated long weights and short weights as

$$w_+ = \sum_{w_i > 0} w_i, w_- = \sum_{w_i < 0} w_i \tag{3.18}$$

Define average return of long assets and short assets by

$$r_+ = \frac{1}{w_+} \sum_{w_i > 0} w_i r_i, r_- = \frac{1}{w_-} \sum_{w_i < 0} w_i r_i \tag{3.19}$$

Then, we have $L = w_+ - w_- > 1$ and $R = w_+ r_+ + w_- r_-$. The analysis of the two-asset portfolio applies directly. We have

$$L^d = \frac{w_+ - w_- + w_+ r_+ - w_- r_-}{1 + w_+ r_+ + w_- r_-} \tag{3.20}$$

and

$$\Delta L = \frac{2w_+ w_- \left(r_+ - r_-\right)}{1 + R} \tag{3.21}$$

Hence, when the weighted average return of long assets is higher than the weighted average return of the short assets, the leverage decreases, and vice versa.

When the number of assets is small, it is possible to pay attention to individual assets. However, when the number of assets is large, grouping them into long assets and short assets is a useful technique, and we shall use it again in later chapters.

3.4 REBALANCING ALPHA

Having gained some understanding of the rebalancing process, the change in portfolio weights, and portfolio leverage, we now formally define rebalancing alpha as the return difference between a fixed-weight portfolio and its buy-and-hold counterpart, which has the same original weights as the fixed-weight portfolio. The return will be the geometric return instead of the arithmetic return, because geometric return determines the wealth of investments. Throughout the rest of the book, we shall use FW and BH to denote fixed-weight and buy-and-hold portfolios, respectively.

Suppose we have M investable assets and an investment horizon of N periods. We assume that in the beginning, both FW and BH portfolios are invested in the assets with portfolio weights: $\mathbf{w} = \left(w_1, \ldots, w_M \right)'$, and the sum of the weights is 1.

We denote return of asset i in period n by r_{in}. For the FW portfolio, the portfolio return is the weighted sum of individual returns with the fixed weights, i.e.

$$r_{FW,n} = w_1 r_{1n} + w_2 r_{2n} + \cdots + w_M r_{Mn} = \sum_{i=1}^{M} w_i r_{in} \tag{3.22}$$

The wealth of the FW portfolio grows by this return over time. The geometric return of the FW portfolio with portfolio rebalancing after N periods is defined by

$$\left(1 + g_{FW} \right)^N = \left(1 + \sum_{i=1}^{M} w_i r_{i1} \right) \cdots \left(1 + \sum_{i=1}^{M} w_i r_{iN} \right) \tag{3.23}$$

On the other hand, the wealth of the BH portfolio grows according to the growth of each individual asset. We have

$$(1+g_{BH})^N = w_1(1+r_{11})\cdots(1+r_{1N})+\cdots+w_M(1+r_{M1})\cdots(1+r_{MN}) \qquad (3.24)$$

Equation 3.24 can be rewritten in terms of the geometric returns of individual assets:

$$(1+g_{BH})^N = w_1(1+g_1)^N+\cdots+w_M(1+g_M)^N \qquad (3.25)$$

We have defined individual returns by

$$(1+g_i)^N = (1+r_{i1})\cdots(1+r_{iN}), i = 1,\ldots, M \qquad (3.26)$$

We define rebalancing alpha as the difference:

$$\alpha = g_{FW} - g_{BH} \qquad (3.27)$$

We make several comments about this definition.

- We have not explicitly solved for g_{FW} and g_{BH} from the equations above. It is easy to do so. However, in most cases, this is not necessary for analysis.

- This definition is used for sample analysis as well as for distributional analysis. The former analysis is mostly concerned with the result of portfolio rebalancing in investment practice. The latter is useful in the theoretical study of portfolio theory. We focus more attention on the former in this book. However, we devote a chapter on the expected wealth of FW and BH portfolios.

3.4.1 Rebalancing Alpha of Asset Allocation Portfolios

We now provide three real-world examples of rebalancing alpha based on asset allocation portfolios. These portfolios are quite simple—they are made up of two assets out of cash, bonds, stocks, and commodities. The return statistics of the four assets are displayed in Table 2.1. In addition, the portfolio weights all start at 50/50. These three examples show that rebalancing alpha can be quite different for different portfolios.

TABLE 3.9 Geometric Return and Volatility of 50/50 Cash/Stock Portfolios

	BH	FW
g	9.04%	8.21%
σ	12.24%	8.95%

The first is a 50/50 cash/stock portfolio. The annual returns are from 1970 to 2014, and the rebalancing frequency is annual. In this chapter, we shall not include transaction costs in the analysis. Table 3.9 shows the geometric returns and volatilities of the FW and BH portfolios. In this case, the BH portfolio's return is 9.04% per year, while the FW portfolio's return is lower at 8.21%. Hence, the rebalancing alpha for this portfolio is negative 83 bps. This might not appear as a big difference. But the difference of compound wealth of the two portfolios is quite significant. Table 3.9 also shows the return volatilities of BH and FW portfolios. The BH portfolio has significantly higher volatility than the FW portfolio.

In Figure 3.1, we plot the cumulative wealth of the FW and BH portfolios, both starting at 1 at the end of 1969. The logarithmic scale is used on the vertical axis. By the end of 2014, the wealth of the BH portfolio is about

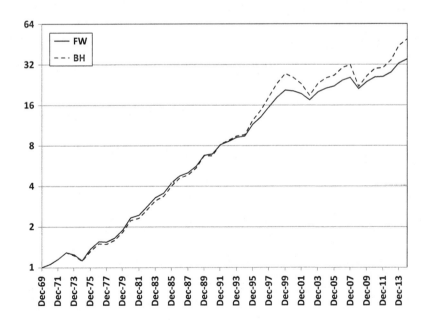

FIGURE 3.1 Cumulative wealth of BH and FW portfolios with 50% in cash and 50% in stocks.

TABLE 3.10   Geometric Return and Volatility
of 50/50 Bond/Stock Portfolios

	BH	FW
g	9.43%	9.49%
σ	11.47%	9.59%

49, while the wealth of the FW portfolio is close to 35. It is noted that the two lines only separated after 1993, and prior to 1993, the wealth of the two portfolios tracked each other closely. In fact, in the first half of the sample, it is the FW portfolio that has a slight edge. This change of fortune highlights the empirical fact that rebalancing alpha depends on the specific period.

The second example is the 50/50 bond/stock portfolios. The results of the BH and FW portfolios are shown in Table 3.10. In this case, the geometric returns are almost the same, with just 6 bps of difference. However, the volatilities are quite different, as in the cash/stock case. Portfolio rebalancing does seem to reduce portfolio risk compared with the BH portfolio. The cumulative wealth of the two portfolios is shown in Figure 3.2. Throughout the early period, the two lines are close to each other. It is only from the late 1990s that the two lines show a large deviation due to higher volatility in stock returns.

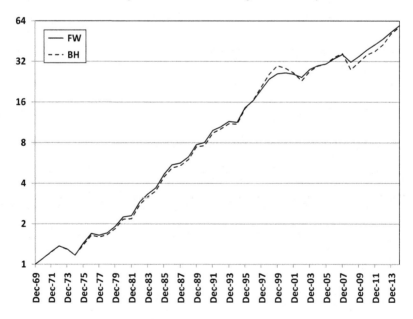

FIGURE 3.2   Cumulative wealth of BH and FW portfolios with 50% in bonds and 50% in stocks.

TABLE 3.11   Geometric Return and Volatility of
50/50 Stock/Commodity Portfolios

	BH	FW
g	9.54%	10.39%
σ	15.80%	14.65%

The last example is the stock/commodity portfolios. The results of the BH and FW portfolios are shown in Table 3.11. Here, the FW portfolio outperforms the BH portfolio significantly, and the rebalancing alpha is 85 bps. Once again, this is achieved by lower volatility for the FW portfolio. The reduction in volatility is not as high as in the previous two cases, since stocks and commodities all have high return volatilities.

Figure 3.3 shows the cumulative wealth of the two portfolios. The terminal wealth of the FW portfolio is at 85.5, while the terminal wealth of the BH portfolio is at 60.5. It is interesting to note that the FW has outperformed the BH portfolio since 1980, with an ever-widening gap between the two lines.

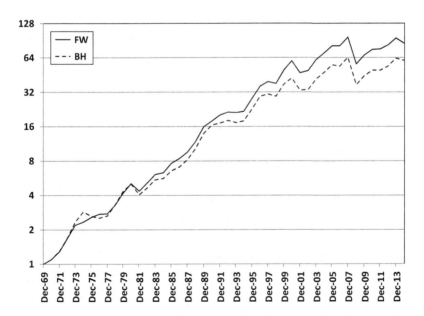

FIGURE 3.3   Cumulative wealth of BH and FW portfolios with 50% in stocks and 50% in commodities.

3.4.2 Periodic Rebalancing versus Threshold Rebalancing

We close this chapter with a brief discussion about threshold rebalancing. So far, we have made an implicit assumption that portfolio rebalancing is carried out regularly and periodically based on the calendar. For instance, we used annual rebalancing in our previous examples. Portfolio rebalancing could be done on a monthly or quarterly basis or at any other regularly scheduled interval. Periodic rebalancing is convenient in practice, and it is mathematically more tractable than another form of portfolio rebalancing that is based not on the calendar but on magnitudes of weight changes. However, this does not have to be the case. Portfolio rebalancing could be irregular. Threshold rebalancing is conditional, and it might be dependent on market-based variables or even proprietary models. However, the most natural condition is based on the magnitude of weight changes. We shall return to this topic in Chapter 12.

Threshold rebalancing is a result of two considerations in portfolio management. The first is that portfolio weights, especially asset allocation weights, often carry with them a permissible range. For instance, a 60/40 asset allocation portfolio with a target weight of 60% in stocks and 40% in bonds might have a permissible range for stocks between 55% and 65% and a corresponding range for bonds between 45% and 35%. An investor could decide to rebalance the portfolio back to 60/40 only when return drift causes the portfolio to cross the boundary of the range. Within the range, investors might consider the difference between the portfolio and the target to be small.

Another motivation for threshold rebalancing is related to the generation of rebalancing alpha. For example, if returns have a stronger tendency to mean reversion when portfolio weights drift further away from the target weights in a long-only portfolio, then a carefully calibrated threshold (or conditional) rebalancing might produce higher rebalancing alpha than periodic (unconditional) rebalancing.

We note that the notations developed in Equations 3.22 through 3.27 are equally applicable to threshold rebalancing. It is obvious that the return of a BH portfolio is unaffected by the method of rebalancing. For an FW portfolio, the return $r_{FW,n}$ is simply its return in period n, but the length of periods differs from one to the next. If calendar-based rebalancing occurs monthly, then threshold rebalancing would occur irregularly, and the period between rebalancings could be a month, two months, or even a year.

PROBLEMS

3.1. Suppose we have two investments, A and C. Investment A returns 100% in Year 1 and −50% in Year 2. Investment C returns 60% in Year 1 and −37.5% in Year 2. Show that (a) two investments have zero cumulative return and a correlation of 1; (b) a fixed-weight portfolio of 50/50 generates a positive return over the two-year period.

3.2. (a) Prove that the product $w_1 w_2$ with $w_1 + w_2 = 1$ is maximized at $w_1 = w_2 = 0.5$; (b) find the portfolio weights that maximize Δw_1 of Equation 3.4 when $r_1 > r_2$; (c) show that the optimal weight w_1 is approximately $w_1 \approx 0.5 - (r_1 - r_2)/8$.

3.3. Prove Equations 3.11 and 3.12.

3.4. Show that the standard deviation of portfolio turnover for a portfolio of two assets, whose expected returns are identical, is

$$\text{std}(T) = \sqrt{1 - \frac{2}{\pi}} |w_1 w_2| \sigma_{1-2}$$

3.5. Estimate the expected turnover of an equal-weight portfolio in stock, bond, and commodity with inputs from Table 3.5.

3.6. Construct (a) an example in which a two-asset long-short portfolio has a positive return, and portfolio leverage increases; (b) an example in which a two-asset long-short portfolio has a negative return, and portfolio leverage decreases.

CHAPTER 4

Volatility Effect and Return Effect

I N Chapter 3, we defined the rebalancing alpha as the difference between the geometric returns of fixed-weight (FW) and buy-and-hold (BH) portfolios. Starting from this chapter, we begin the analysis of rebalancing alpha. The most common term used in the study of portfolio rebalancing is *diversification return*. As we shall see shortly, for a long-only portfolio, the diversification return is always non-negative. This is often taken as the proof that rebalancing alpha is positive. In other words, portfolio rebalancing adds value for long-only portfolios.

This claim is false. The empirical examples we have provided in the previous chapters definitively show that various outcomes are possible for rebalancing alpha of long-only portfolios. Theoretically, the claim is false for a very simple reason: the diversification return does not equate to rebalancing alpha. It is only a part of rebalancing alpha, which can be identified as the volatility effect of portfolio rebalancing. There is a second effect, the return effect of portfolio rebalancing, which often acts counter to the volatility effect. And the rebalancing alpha is the net of the two effects.

To proceed further, we shall define the two effects in this chapter and study them separately in the next few chapters. In most cases, we must analyze both effects to get an answer for rebalancing alpha. Only in certain special cases is rebalancing alpha determined by a single effect. We present these special cases in this chapter. They are not only of mathematical interest but also of practical relevance, for they offer insights into the relative importance of volatility and return effects in different cases.

4.1 DEFINITIONS OF TWO EFFECTS

For ease of reference, we rewrite the definition of returns of FW and BH portfolios. For the FW portfolio, we have

$$\left(1+g_{FW}\right)^N = \left(1+\sum_{i=1}^{M}w_i r_{i1}\right)\cdots\left(1+\sum_{i=1}^{M}w_i r_{iN}\right) \tag{4.1}$$

For the BH portfolio, the return is given by

$$\left(1+g_{BH}\right)^N = w_1\left(1+g_1\right)^N+\cdots+w_M\left(1+g_M\right)^N \tag{4.2}$$

And $g_1,\cdots,g_M$ are the geometric returns of individual assets. The rebalancing alpha is $\alpha = g_{FW} - g_{BH}$.

In most cases, it is rather difficult to make a direct comparison between g_{FW} and g_{BH}. The way to overcome this difficulty is to introduce a third "return" term and compare both g_{FW} and g_{BH} with this term. This "return" term is the average of geometric returns. We have

$$\overline{g} = w_1 g_1 + \cdots + w_M g_M = \sum_{i=1}^{M}w_i g_i \tag{4.3}$$

It is important to note that $\overline{g}$ is a fictitious "return" in the sense that it is not the geometric return of any actual portfolio, and certainly not the geometric return of the BH portfolio. We define volatility effect as the difference between the geometric returns of the FW portfolio and $\overline{g}$:

$$e_v = g_{FW} - \overline{g} \tag{4.4}$$

The other half of rebalancing alpha is return effect, defined as

$$e_r = g_{BH} - \overline{g} \tag{4.5}$$

Then, we have

$$\alpha = g_{FW} - g_{BH} = e_v - e_r \tag{4.6}$$

Rebalancing alpha is the difference between the volatility and return effects. The reasons for the names of these two effects will be made clear later. For the time being, we note that the volatility effect is mainly caused by asset return volatility, and the return effect is due to dispersion of average returns of individual assets.

Example 4.1: When geometric returns of individual assets are the same, by Equation 4.2, we have $g_{BH} = \bar{g}$. In this case, the average geometric return is the same as the return of the BH portfolio, and the return effect is zero. Thus, the rebalancing alpha is just volatility effect, that is, $\alpha = e_v$.

- If we look back at Example 3.1, in which two investments, A and B, both have zero cumulative return after two years, we note that $g_{BH} = \bar{g} = 0$ for any BH portfolios. In that example, the FW portfolio has a positive return, and the rebalancing alpha is positive.
- The fact that the rebalancing alpha is positive in Example 3.1 is not accidental. We shall prove that rebalancing alpha is always positive for long-only portfolios when geometric returns are the same for individual assets.

Example 4.2: As an empirical example, we refer to the 50/50 bond/stock portfolios in Chapter 3. From Tables 2.1 and 3.10, we gather the following returns in Table 4.1. The average of geometric returns is 9.03%. Thus, even though the rebalancing alpha is negligibly small, both volatility and return effects are positive.

- We find the same in the examples of 50/50 cash/stock portfolios and 50/50 stock/commodity portfolios. In other words, the volatility and return effects are positive.

TABLE 4.1   Returns, Effects, and Rebalancing Alpha of 50/50 Bond/Stock Portfolios

g_{BH}	g_{FW}	$\bar{g}$	e_v	e_r	α
9.43%	9.49%	9.03%	0.46%	0.40%	−0.06%

4.2 POSITIVE RETURN EFFECT OF LONG-ONLY PORTFOLIOS

We now prove mathematically that both effects are non-negative for long-only portfolios. To prove the case for the return effect, we shall use Jensen's inequality for convex functions. This result is connected to arithmetic mean–geometric mean (AM-GM) inequality, and it is useful in proving inequality for functions instead of number sequences.

4.2.1 Jensen's Inequality

Suppose $f(x)$ is a real-valued function, and its second derivative is non-negative $f''(x) \geq 0$. Then, we have

$$f\left(\sum_{i=1}^{M} w_i x_i\right) \leq \sum_{i=1}^{M} w_i f(x_i) \tag{4.7}$$

The variables $x_i, i = 1, \cdots, M$ are arbitrary, the weights are non-negative, and $\sum_{i=1}^{M} w_i = 1$.

Equation 4.7 states that for convex functions, the weighted average of function-values is greater than or equal to the function of the weighted average. Geometrically, when $M = 2$, it states that the line segment connecting two points on the curve $y = f(x)$ lies atop of the curve between those two points.

- When the second derivation is always positive, then the function $y = f(x)$ is strictly convex. For strictly convex functions, the equality in Equation 4.7 is true in general only if $x_1 = x_2 = \cdots = x_M$.

- Jensen's inequality requires that the weights are non-negative. Therefore, it is mostly suitable for analyzing long-only portfolios. However, when some of the weights are negative, such as in the case of long-short portfolios, Jensen's inequality becomes undetermined. Most astonishingly, in some cases, the reverse is true (see Problem 4.2).

Example 4.3: If $f(x) = (1+x)^N$ with $N \geq 2$ and $x > -1$, then the second derivative $f''(x) = N(N-1)(1+x)^{N-2} > 0$. Applying Jensen's inequality yields

$$\left(1+\sum_{i=1}^{M}w_i x_i\right)^N \le \sum_{i=1}^{M}w_i\left(1+x_i\right)^N \tag{4.8}$$

4.2.2 Return Effect of Long-Only Portfolios

Taking x_i as the geometric return g_i of an individual asset, we have

$$\left(1+\sum_{i=1}^{M}w_i g_i\right)^N \le \sum_{i=1}^{M}w_i\left(1+g_i\right)^N \tag{4.9}$$

According to Equation 4.2, the right-hand side is $\left(1+g_{BH}\right)^N$. The left-hand side is, of course, $\left(1+\bar{g}\right)^N$. Hence, inequality Equation 4.9 proves that $g_{BH} \ge \bar{g}$. Therefore, for long-only portfolios, we have a non-negative return effect:

$$e_r = g_{BH} - \bar{g} \ge 0 \tag{4.10}$$

Furthermore, the return effect is zero only when the geometric returns of all assets are the same, since $f(x) = (1+x)^N$ is strictly positive. In other words, the condition for zero return effect is when

$$g_1 = g_2 = \cdots = g_M \tag{4.11}$$

In this special case, the return of the BH portfolio equals the return of all individual assets. In addition, if the volatility effect is positive, then the rebalancing alpha will be positive for this special case. Intuitively, portfolio rebalancing of buying losers and selling winners in a long-only portfolio would generate positive alpha, since interim winners and losers eventually revert to the same return. The simple example given at the beginning of Chapter 3 confirms this intuition.

4.3 POSITIVE VOLATILITY EFFECT OF LONG-ONLY PORTFOLIOS

The goal of this section is to prove that for any long-only portfolio, the volatility effect is always non-negative. In other words, the geometric return of an FW portfolio is greater than or equal to the weighted average of

geometric returns. Mathematically, we show that for any long-only portfolio weights w_i, any number of assets M, and any number of periods N,
we have

$$\left(1+\sum_{i=1}^{M}w_i r_{i1}\right)\cdots\left(1+\sum_{i=1}^{M}w_i r_{iN}\right)\geq\left(1+\sum_{i=1}^{M}w_i g_i\right)^N \tag{4.12}$$

The general inequality Equation 4.12 can be proved. To gain insight into
the volatility effect, we first consider a simple case with two assets and two
investment periods.

4.3.1 Cauchy's Inequality

The analysis of volatility effect for long-only portfolios relies on Cauchy's
inequality and its generalization. Suppose $a_i, b_i, i = 1, \cdots, N$ are real positive
numbers. Then

$$\left(\sum_{i=1}^{N}\sqrt{a_i b_i}\right)^2 \leq \left(\sum_{i=1}^{N}a_i\right)\left(\sum_{i=1}^{N}b_i\right) \tag{4.13}$$

- We have rewritten Cauchy's inequality in a slightly different form.
  It is easier to use in our analysis that follows.

- The inequality Equation 4.13 has many interpretations. If we con-
  sider $\mathbf{a} = \left(\sqrt{a_1},\cdots,\sqrt{a_N}\right)'$ and $\mathbf{b} = \left(\sqrt{b_1},\cdots,\sqrt{b_N}\right)'$ as two vectors in
  N-dimensional space, then it states that the inner product of the two
  vectors is less than or equal to the product of the lengths of the two
  vectors; that is, $|\mathbf{a} \cdot \mathbf{b}| \leq \|\mathbf{a}\|\|\mathbf{b}\|$. In addition, it is an equality only if the
  two vectors are parallel; that is, $\mathbf{a} = k\mathbf{b}$.

4.3.2 A Two-Asset Two-Period Case

We first consider this simplest case to gain some intuition. Here, we
can proceed directly without using Cauchy's inequality. According to
Equation 4.1, we have

$$\left(1+g_{FW}\right)^2 = \left(1+w_1 r_{11} + w_2 r_{21}\right)\left(1+w_1 r_{12} + w_2 r_{22}\right)$$

$$= \left[w_1\left(1+r_{11}\right) + w_2\left(1+r_{21}\right)\right]\left[w_1\left(1+r_{12}\right) + w_2\left(1+r_{22}\right)\right] \tag{4.14}$$

Recall that the first subscript denotes the asset and the second subscript denotes the period. Multiplying out the right-hand side, we have

$$\left(1+g_{FW}\right)^2 = w_1^2\left(1+r_{11}\right)\left(1+r_{12}\right) + w_2^2\left(1+r_{21}\right)\left(1+r_{22}\right)$$
$$+ w_1 w_2\left[\left(1+r_{11}\right)\left(1+r_{22}\right) + \left(1+r_{21}\right)\left(1+r_{12}\right)\right]$$
(4.15)

We recognize that the first two terms contain cumulative returns of individual assets. Hence,

$$\left(1+g_{FW}\right)^2 = w_1^2\left(1+g_1\right)^2 + w_2^2\left(1+g_2\right)^2$$
$$+ w_1 w_2\left[\left(1+r_{11}\right)\left(1+r_{22}\right) + \left(1+r_{21}\right)\left(1+r_{12}\right)\right]$$
(4.16)

The last two terms consist of returns of different assets from different periods. We use AM-GM inequality to get

$$\left(1+r_{11}\right)\left(1+r_{22}\right) + \left(1+r_{21}\right)\left(1+r_{12}\right) \geq 2\sqrt{\left(1+r_{11}\right)\left(1+r_{12}\right)\left(1+r_{21}\right)\left(1+r_{22}\right)}$$
$$= 2\left(1+g_1\right)\left(1+g_2\right)$$
(4.17)

In the last step, we simply used the definition of geometric returns of two assets. Using Equations 4.17 in 4.16, we have

$$\left(1+g_{FW}\right)^2 \geq w_1^2\left(1+g_1\right)^2 + w_2^2\left(1+g_2\right)^2 + 2w_1 w_2\left(1+g_1\right)\left(1+g_2\right)$$
$$= \left[w_1\left(1+g_1\right) + w_2\left(1+g_2\right)\right]^2 = \left(1+w_1 g_1 + w_2 g_2\right)^2$$
(4.18)

This leads to

$$g_{FW} \geq \overline{g} = w_1 g_1 + w_2 g_2 \tag{4.19}$$

Hence, we have proved that in this case, the geometric return of the FW portfolio is greater than or equal to the average of geometric return. As a result, the volatility effect is non-negative.

It is interesting to investigate when the volatility effect is zero in this simple case. It occurs when Equation 4.17 is an equality:

$$\left(1+r_{11}\right)\left(1+r_{22}\right) = \left(1+r_{21}\right)\left(1+r_{12}\right) \tag{4.20}$$

The left-hand side is the cumulative return of investing in Asset 1 in Period 1 and then switching to Asset 2 in Period 2. The right-hand side is the cumulative return of investing in Asset 2 in Period 1 and then switching to Asset 1 in Period 2. Both investments involve switching assets after Period 1. In a sense, this is similar to portfolio rebalancing, in which we swap one asset for another asset. Equation 4.20 makes it clear that when the returns are indifferent to asset swapping, the volatility effect of portfolio rebalancing is zero.

Equation 4.20 can also be rewritten as the ratios of returns. We have

$$\frac{(1+r_{11})}{(1+r_{21})} = \frac{(1+r_{12})}{(1+r_{22})} = k \tag{4.21}$$

The left-hand side is the ratio of asset return (plus one) in Period 1, and the right-hand side is the same ratio in Period 2. The ratio could be any positive number. However, as long as the ratio is the same for two periods, the volatility effect of rebalancing is zero. If the ratio is 1, the two assets have identical returns in both periods. There is no rebalancing in this case and hence, no volatility effect. If the ratio is greater than 1, then Asset 1 outperforms Asset 2 in both periods. On the other hand, if the ratio is less than 1, Asset 1 underperforms Asset 2 in both periods. There will be portfolio rebalancing in both cases. When Equation 4.21 holds, there is no volatility effect. There will be return effects, however, since the two assets have different returns. In these cases, the rebalancing alpha will be negative.

- Equation 4.21 is implied by the equality condition for Cauchy's inequality; that is, the return vector of Asset 1 is proportional to the return vector of Asset 2.

Example 4.4: Suppose we have two assets whose returns in two periods are shown in Table 4.2. Then, it is noted that the condition in Equation 4.21 holds, and the ratio is 1.2. Asset 1 outperforms Asset 2 in both periods. The geometric returns are −5.13% and −20.94%, respectively. Suppose we have a 40/60 portfolio, with $w_1 = 40\%$ and

TABLE 4.2   Returns of Two Assets in Two Periods

$r_{11} = 50\%$	$r_{12} = -40\%$	$g_1 = -5.13\%$
$r_{21} = 25\%$	$r_{22} = -50\%$	$g_2 = -20.94\%$

$w_2 = 60\%$. Then, $\overline{g} = -14.62\%$. The return of an FW portfolio is 35% in Period 1 and −46% in Period 2. These yield a geometric return $g_{FW} = -14.62\%$, which is the same as $\overline{g}$. There is no volatility effect. It is easy to verify that the return effect is nonzero. Therefore, the rebalancing alpha is negative.

- Our analysis reveals that volatility effect, while in general non-negative, could be zero in special cases. The case in which all assets are the same is one such special case. However, even when asset returns are not identical, the volatility return can still vanish.
- When the condition in Equation 4.20 is met, any portfolio, whether long-only or long-short, would have zero volatility effect. This is true for general cases.
- As we shall see shortly in the next chapter, diversification return, which serves as an approximation of volatility effect, is positive unless all assets are identical. Hence, this reflects a difference between volatility effect and diversification return.

4.3.3 An M-Asset Two-Period Case

Before discussing the general result, we investigate another special case with M assets and two periods. The Cauchy's inequality that clinched this case can be succinctly written as

$$\left(a_1 + a_2 + \cdots + a_M\right)\left(b_1 + b_2 + \cdots + b_M\right) \geq \left(\sqrt{a_1 b_1} + \sqrt{a_2 b_2} + \cdots + \sqrt{a_M b_M}\right)^2 \quad (4.22)$$

Let $a_i = w_i\left(1 + r_{i1}\right)$, $b_i = w_i\left(1 + r_{i2}\right)$. In other words, a_is are returns in the first period, and b_is are returns in the second period. We have

$$\left(1 + w_1 r_{11} + \cdots + w_M r_{M1}\right)\left(1 + w_1 r_{12} + \cdots + w_M r_{M2}\right) \geq \left(1 + w_1 g_1 + \cdots + w_M g_M\right)^2 \quad (4.23)$$

This is equivalent to $g_{FW} \geq \overline{g}$.

The equality in Equation 4.23 occurs when we have

$$\frac{\left(1 + r_{12}\right)}{\left(1 + r_{11}\right)} = \frac{\left(1 + r_{22}\right)}{\left(1 + r_{21}\right)} = \cdots = \frac{\left(1 + r_{M2}\right)}{\left(1 + r_{M1}\right)} = k \quad (4.24)$$

This is the condition of equality for Cauchy's inequality. The ratios of return in Period 1 to return in Period 2 are constant for all assets. There are two variations of this condition. First, we derive the following pairwise conditions:

$$\left(1+r_{i1}\right)\left(1+r_{j2}\right)=\left(1+r_{j1}\right)\left(1+r_{i2}\right), \text{ for } i, j =1,\cdots, M \qquad (4.25)$$

Similarly to Equation 4.20, each equation describes an equivalency of investment results in a pair of assets when the order of investment is exchanged. Since there are M assets, there are $M(M-1)/2$ pairs. However, only $(M-1)$ equations in Equation 4.25 are independent.

Second, from Equation 4.25, the condition can also be rewritten for the ratio of returns between pairs of assets for two different periods:

$$\frac{\left(1+r_{i1}\right)}{\left(1+r_{j1}\right)}=\frac{\left(1+r_{i2}\right)}{\left(1+r_{j2}\right)}, \text{ for } i, j =1,\cdots, M \qquad (4.26)$$

Example 4.5: Suppose we have three assets, whose returns in two periods are shown in Table 4.3. We have

$$\frac{1.5}{0.6}=\frac{1.25}{0.5}=\frac{1}{0.4}=k=2.5$$

Thus, any portfolio comprised of these three assets, no matter what the portfolio weights are, will have zero volatility effect.

4.3.4 The General Case

For the general case with M assets and N periods, we prove Equation 4.12 by using an extension of Cauchy's inequality from two vectors to M vectors. We have, for non-negative numbers a_{ij},

TABLE 4.3   Returns of Three Assets for Two Periods

	Period 1	Period 2
Asset 1	$r_{11} = 50\%$	$r_{12} = -40\%$
Asset 2	$r_{21} = 25\%$	$r_{22} = -50\%$
Asset 3	$r_{31} = 0\%$	$r_{32} = -60\%$

$$\left(a_{11}+a_{21}+\cdots+a_{M1}\right)\cdots\left(a_{1N}+a_{2N}+\cdots+a_{MN}\right)$$

$$\geq \left(\sqrt[N]{a_{11}a_{12}\cdots a_{1N}}+\cdots+\sqrt[N]{a_{M1}a_{M2}\cdots a_{MN}}\right)^N \qquad (4.27)$$

We have resorted to using two subscripts, one for asset and the other for period. It is obvious that when $N=2$, Equation 4.27 becomes the traditional Cauchy's inequality, which is used in the analysis of the case of M assets and two periods. When $N>2$, Equation 4.27 can be proved by mathematical induction. We leave it as an exercise.

Letting $a_{ij}=w_i\left(1+r_{ij}\right)$ in Equation 4.27 leads us directly to Equation 4.12. Thus, we have proved that for long-only portfolios, the volatility effect is always non-negative.

The volatility effect is zero when Equation 4.27 becomes an equality. This occurs when N vectors of returns are proportional to each other, similarly to Equation 4.24. Instead of two periods, we now have N periods. The equality condition can be expressed in terms of the following $N-1$ conditions. We have

$$\begin{cases} \dfrac{\left(1+r_{12}\right)}{\left(1+r_{11}\right)}=\dfrac{\left(1+r_{22}\right)}{\left(1+r_{21}\right)}=\cdots=\dfrac{\left(1+r_{M2}\right)}{\left(1+r_{M1}\right)}=k_2, \\[2mm] \dfrac{\left(1+r_{13}\right)}{\left(1+r_{11}\right)}=\dfrac{\left(1+r_{23}\right)}{\left(1+r_{21}\right)}=\cdots=\dfrac{\left(1+r_{M3}\right)}{\left(1+r_{M1}\right)}=k_3, \\[2mm] \qquad\qquad\qquad\qquad \vdots \\[2mm] \dfrac{\left(1+r_{1N}\right)}{\left(1+r_{11}\right)}=\dfrac{\left(1+r_{2N}\right)}{\left(1+r_{21}\right)}=\cdots=\dfrac{\left(1+r_{MN}\right)}{\left(1+r_{M1}\right)}=k_N \end{cases} \qquad (4.28)$$

Denoting return plus one of each period by

$$\mathbf{R}_j=\left[\left(1+r_{1j}\right),\left(1+r_{2j}\right),\cdots,\left(1+r_{Mj}\right)\right], j=1,\cdots,N \qquad (4.29)$$

Then, conditions in Equation 4.28 are

$$\mathbf{R}_j=k_j\mathbf{R}_1, j=2,\cdots,N \qquad (4.30)$$

Of course, if all return vectors are of constant proportional of return vector of first period, they are proportional to one another

From Equation 4.28 or 4.30, we derive

$$\left(1+r_{il}\right)\left(1+r_{jk}\right)=\left(1+r_{jl}\right)\left(1+r_{ik}\right),$$

$$\text{for } i,j=1,\cdots,M \text{ and } l,k=1,\cdots,N$$

(4.31)

Equation 4.31 states that for any pair of assets and any two periods, investment returns are the same when the two assets are swapped across time.

- When Equation 4.30 is met, the volatility effect becomes zero. As we have shown previously, the return effect is always non-negative. As a result, the rebalancing alpha, which is volatility effect minus return effect, is likely to be negative. We shall present a direct proof later.

- When the return vectors from difference periods are linearly and positively proportional to each other, as indicated by Equation 4.30, it implies that the cross-sectional correlations of those return vectors are a perfect 1; that is, $\text{corr}\left(\mathbf{R}_i,\mathbf{R}_j\right)=1$ for all $i,j=1,\cdots,N$. These perfect correlations can be thought of as perfect cross-sectional momentum. It is not entirely surprising that volatility effect vanishes, and rebalancing alpha is likely to be negative.

Example 4.6: We extend Example 4.5 by one period. Suppose that in Period 3, the returns are 20%, 0%, and −20%, respectively. Then, the ratio of returns in Period 1 to returns in Period 3 is

$$\frac{1.5}{1.2}=\frac{1.25}{1}=\frac{1}{0.8}=1.25$$

Thus, any portfolio comprised of these three assets, no matter what the portfolio weights are, will have zero volatility effect over these three periods (Table 4.4).

TABLE 4.4 Returns of Three Assets for Two Periods

	Period 1	Period 2	Period 3
Asset 1	$r_{11}=50\%$	$r_{12}=-40\%$	$r_{13}=20\%$
Asset 2	$r_{21}=25\%$	$r_{22}=-50\%$	$r_{23}=0\%$
Asset 3	$r_{31}=0\%$	$r_{32}=-60\%$	$r_{32}=-20\%$

4.4 CASES OF POSITIVE AND NEGATIVE REBALANCING ALPHAS

We have proved separately that both volatility and return effect of long-only portfolios are non-negative. The results are affirmative and will be useful for the analysis of these two effects in later chapters. However, these results in general do not lead to definitive answers about rebalancing alpha, which represents the net effect of the two. Only in certain special cases can we tell whether rebalancing alpha is either positive or negative.

We have encountered these special cases. They are the ones in which either return effect or volatility effect is zero. In this section, we revisit them by studying rebalancing alpha directly, thus skipping the analysis of two individual effects. Even though the separate analysis is essential for much of the empirical and analytical study of rebalancing alpha, a direct attack is quite useful in some circumstances.

4.4.1 The Case of Positive Rebalancing Alpha

The return effect is zero when the geometric returns of all assets are the same, that is, $g_1 = g_2 = \cdots = g_M$. In this case, the return of the BH portfolio is just the same $g_{\text{BH}} = g_1$, and we now prove directly that $g_{\text{FW}} \geq g_{\text{BH}}$.

According to Equation 4.1, we have

$$\left(1 + g_{\text{FW}}\right)^N = \left[\sum_{i=1}^{M} w_i \left(1 + r_{i1}\right)\right] \cdots \left[\sum_{i=1}^{M} w_i \left(1 + r_{iN}\right)\right] \qquad (4.32)$$

Each bracket is a weighted sum (with weight w_i) of positive numbers. We apply the general AM-GM inequality to each bracket. For example, the first is the return of the first period; we have

$$w_1 \left(1 + r_{11}\right) + w_2 \left(1 + r_{21}\right) + \cdots + w_M \left(1 + r_{M1}\right)$$

$$\geq \left(1 + r_{11}\right)^{w_1} \left(1 + r_{21}\right)^{w_2} \cdots \left(1 + r_{M1}\right)^{w_M} \qquad (4.33)$$

Collecting all the terms, we have

$$
\begin{aligned}
\left(1+g_{FW}\right)^{N} &\geq \left[\left(1+r_{11}\right)\cdots\left(1+r_{1N}\right)\right]^{w_{1}}\left[\left(1+r_{21}\right)\cdots\left(1+r_{2N}\right)\right]^{w_{2}} \\
&\quad \cdots\left[\left(1+r_{M1}\right)\cdots\left(1+r_{MN}\right)\right]^{w_{M}} \\
&= \left[\left(1+g_{1}\right)\right]^{Nw_{1}}\left[\left(1+g_{2}\right)\right]^{Nw_{2}}\cdots\left[\left(1+g_{M}\right)\right]^{Nw_{M}} \\
&= \left[\left(1+g_{1}\right)\right]^{N\left(w_{1}+\cdots+w_{M}\right)} = \left(1+g_{1}\right)^{N}
\end{aligned}
\tag{4.34}
$$

It follows that $g_{FW} \geq g_{BH}$.

The two returns are equal only when the GM-AM inequality becomes an equation for returns in every period. This in turn requires that returns of all assets are the same. Hence, when the geometric returns of assets are the same, unless returns of all assets are the same in every period, we have $g_{FW} > g_{BH}$ and $\alpha = g_{FW} - g_{BH} > 0$.

4.4.2 The Case of Negative Rebalancing Alpha

The volatility effect is zero when return vectors $\mathbf{R}_{j}, j = 1,\cdots,N$ in Equation 4.29 are proportional to one another. We use this condition to simplify the return of FW and BH portfolios. Specifically, for returns in any period, we have $\left(1+r_{ij}\right) = k_{j}\left(1+r_{i1}\right)$. For the FW portfolio, the return in period j is then a multiple of return in Period 1. We have

$$
\left(1+g_{FW}\right)^{N} = k_{2}k_{3}\cdots k_{N}\left[\sum_{i=1}^{M}w_{i}\left(1+r_{i1}\right)\right]^{N}
\tag{4.35}
$$

For the BH portfolio, we have

$$
\begin{aligned}
\left(1+g_{BH}\right)^{N} &= w_{1}\left(1+r_{11}\right)\left(1+r_{12}\right)\cdots\left(1+r_{1N}\right) + \cdots \\
&\quad + w_{M}\left(1+r_{M1}\right)\left(1+r_{M2}\right)\cdots\left(1+r_{MN}\right) \\
&= w_{1}\left(1+r_{11}\right)^{N}k_{2}\cdots k_{N} + \cdots + w_{M}\left(1+r_{M1}\right)^{N}k_{2}\cdots k_{N} \\
&= k_{2}\cdots k_{N}\left[w_{1}\left(1+r_{11}\right)^{N} + \cdots + w_{M}\left(1+r_{M1}\right)^{N}\right]
\end{aligned}
\tag{4.36}
$$

Comparing Equations 4.35 and 4.36, and using Jensen's inequality for the two terms in the brackets, we conclude that $g_{FW} \leq g_{BH}$ in this case. So, the rebalancing alpha is less than or equal to zero.

It is zero only if the Jensen's inequality becomes an equality. This is true if returns of all assets in Period 1 are the same. Since returns of other periods are proportional to returns of Period 1, returns in other periods must be the same for all assets too. We note that this is the same condition for zero rebalancing alpha in the previous section. Other than in this special case, we have $\alpha < 0$ when $\mathbf{R}_j, j = 1, \cdots, N$ are proportional to one another.

4.5 TWO-ASSET LONG-SHORT PORTFOLIOS

So far, the theoretical results concerning both volatility and return effects are directed to long-only portfolios. Many standard mathematical results can be readily applied to the case in which all portfolio weights are positive. In contrast, it is hard to apply the mathematical results, such as Jensen's inequality, to long-short portfolios that involve both positive and negative weights.

One should suspect that for long-short portfolios, both volatility and return effects might take on different signs as opposed to long-only portfolios. As discussed previously, rebalancing a long-short portfolio often leads to buying high and selling low, which is the opposite of buying low and selling high as in a long-only portfolio. It is natural to ask whether volatility and return effects should be negative for long-short portfolios.

While this is questionable and probably unlikely in general, we prove that it is the case for long-short portfolios with just two assets. The proof is attainable for two-asset long-short portfolios, because mathematical inequalities used previously can be recast conveniently for this special case.

4.5.1 Negative Return Effect of Two-Asset Long-Short Portfolios

Without loss of generality, we assume that the weight of Asset 1 is negative and the weight of Asset 2 is positive; that is, $w_1 < 0, w_2 > 0$. The return of the BH portfolio is given as

$$\left(1 + g_{BH}\right)^N = w_1 \left(1 + g_1\right)^N + w_2 \left(1 + g_2\right)^N \tag{4.37}$$

We compare g_{BH} to the weighted average $\bar{g} = w_1 g_1 + w_2 g_2$. Since the weight w_1 is negative, $\bar{g}$ is not a conventional average. The key to proving that $g_{BH} \leq \bar{g}$ is to use Jensen's inequality by regrouping terms so that they all have positive "weights."

Consider the sum

$$\left(1+\bar{g}\right)^N - w_1 \left(1+g_1\right)^N = \left(1-w_1\right) \left[\frac{1}{1-w_1} \left(1+\bar{g}\right)^N + \frac{\left(-w_1\right)}{1-w_1} \left(1+g_1\right)^N \right] \quad (4.38)$$

The terms inside the brackets are now a weighted sum of power functions with positive weights. By Jensen's inequality, we have

$$\frac{1}{1-w_1} \left(1+\bar{g}\right)^N + \frac{\left(-w_1\right)}{1-w_1} \left(1+g_1\right)^N \geq \left[1 + \frac{1}{1-w_1} \bar{g} + \frac{\left(-w_1\right)}{1-w_1} g_1 \right]^N \quad (4.39)$$

And the weighted average on the left-hand side yields

$$\frac{1}{1-w_1} \bar{g} + \frac{\left(-w_1\right)}{1-w_1} g_1 = \frac{w_1 g_1 + w_2 g_2 - w_1 g_1}{1-w_1} = \frac{w_2 g_2}{w_2} = g_2 \quad (4.40)$$

Combining these steps together, we have

$$\left(1+\bar{g}\right)^N - w_1 \left(1+g_1\right)^N \geq w_2 \left(1+g_2\right)^N \quad (4.41)$$

Inequality Equations 4.41 and 4.37 together prove that for a two-asset long-short portfolio, we have $g_{BH} \leq \bar{g}$. The return effect is therefore negative unless the two returns are the same.

- The proof for two-asset long-short portfolios relies on the converse of Jensen's inequality, which is true under special conditions.

- The proof can be extended to other long-short portfolios with more than two assets in which only one of the weights is positive and the rest of the weights are negative. For instance, a portfolio that is long stocks and short both bonds and commodities would fit this description. We leave this proof as an exercise.

4.5.2 Negative Volatility Effect of Two-Asset Long-Short Portfolios

The volatility effect of long-only portfolios is always non-negative. We now show that the opposite is true for two-asset long-short portfolios. Together with the result in the previous section, we conclude that two-asset long-short portfolios reverse the signs of both return and volatility effects.

We first prove the result when there are just two periods, that is, $N = 2$, because we have dealt with this case before in Section 4.3. From Equations 4.16 and 4.17, we have

$$\left(1 + g_{FW}\right)^2 \leq w_1^2 \left(1 + g_1\right)^2 + w_2^2 \left(1 + g_2\right)^2 + 2 w_1 w_2 \left(1 + g_1\right)\left(1 + g_2\right)$$

$$= \left[w_1 \left(1 + g_1\right) + w_2 \left(1 + g_2\right)\right]^2 = \left(1 + w_1 g_1 + w_2 g_2\right)^2 \tag{4.42}$$

The inequality in Equation 4.42 takes an opposite direction to that in Equation 4.18, because now $w_1 w_2 < 0$ in a long-short portfolio. Hence, we have $g_{FW} \leq w_1 g_1 + w_2 g_2$.

Unfortunately, this proof cannot be extended to cases in which the number of periods is greater than two, because products of weights will be of mixed signs. For instance, with $N = 3$, we have $w_1^2 w_2 > 0$ and $w_1 w_2^2 < 0$, assuming $w_1 < 0$, $w_2 > 0$.

For cases with M periods, we have

$$\left(1 + g_{FW}\right)^N = \left(1 + w_1 r_{11} + w_2 r_{21}\right) \cdots \left(1 + w_1 r_{1N} + w_2 r_{2N}\right)$$

$$= \left[w_1 \left(1 + r_{11}\right) + w_2 \left(1 + r_{21}\right)\right] \cdots \left[w_1 \left(1 + r_{1N}\right) + w_2 \left(1 + r_{2N}\right)\right] \tag{4.43}$$

We write, for $i = 1, \cdots, N$

$$a_i = \left(b_i - c_i\right)$$

$$b_i = w_2 \left(1 + r_{2i}\right) = a_i + c_i \tag{4.44}$$

$$c_i = -w_1 \left(1 + r_{1i}\right)$$

With $w_1 < 0$, $w_2 > 0$, we have $b_i > 0$, $c_i > 0$ and $a_i > 0$, provided that the portfolio is always solvent. Equation 4.43 can be rewritten as

$$1 + g_{FW} = \sqrt[N]{a_1 \cdots a_N} \tag{4.45}$$

We now use a special case of inequality Equation 4.27 to arrive at

$$\sqrt[N]{a_1 \cdots a_N} + \sqrt[N]{c_1 \cdots c_N} \leq \sqrt[N]{(a_1 + c_1) \cdots (a_N + c_N)} = \sqrt[N]{b_1 \cdots b_N} \quad (4.46)$$

Substituting Equation 4.44 into Equation 4.46, we have

$$1 + g_{FW} - w_1(1 + g_1) \leq w_2(1 + g_2) \quad (4.47)$$

Hence, $g_{FW} \leq w_1 g_1 + w_2 g_2$, which proves the negativity of volatility effect.

- Similarly, this proof can be extended to other long-short portfolios with more than two assets in which only one of the weights is positive and the rest of the weights are negative. We leave this proof as an exercise.

PROBLEMS

4.1. When $f(x) = (1+x)^2$, prove Jensen's inequality directly.

4.2. When $f(x) = (1+x)^2$ and there are just two points: x_1 and x_2. However, instead of non-negative weights, we have $w_1 < 0$ and $w_2 = 1 - w_1 > 0$. Prove that the opposite of Jensen's inequality is true.

4.3. Carry out the calculations in Example 4.4.

4.4. Prove the inequality Equation 4.22.

4.5. Prove generalized Cauchy's inequality Equation 4.27 by mathematical induction.

4.6. Verify that volatility effects are equal to zero in both Examples 4.5 and 4.6.

4.7. For long-short portfolios with M assets, we have $w_i < 0, i = 1, \cdots, M-1$ and $w_M > 0$. Prove that the converse of Equation 4.9 is true; that is,

$$\left(1 + \sum_{i=1}^{M} w_i g_i\right)^N \geq \sum_{i=1}^{M} w_i \left(1 + g_i\right)^N$$

4.8. For long-short portfolios with M assets, we have $w_i < 0, i = 1, \cdots, M-1$ and $w_M > 0$. Prove that the volatility effect is non-positive.

CHAPTER 5

Analysis of Volatility Effect

THIS CHAPTER IS DEVOTED to the analysis of volatility effect. In the previous chapter, we have defined volatility effect of portfolio rebalancing as the difference between the return of a fixed-weight (FW) portfolio and the weighted average of geometric means of individual assets in the portfolio. In addition, we have shown that for long-only portfolios, the volatility effect is always non-negative. While this is a strong mathematical result, it does not give us the actual size of the volatility effect. That is the task in this chapter. We shall derive approximations of volatility effect using volatilities of FW portfolio and individual assets.

5.1 "DIVERSIFICATION RETURN"

The derivation of "diversification return," which is simply an approximation of volatility effect, is based on the approximate relationship between arithmetic and geometric means. It is Equation 2.18 in Chapter 2:

$$g \approx \mu - \frac{\sigma^2}{2} \tag{5.1}$$

We apply it to M individual assets:

$$g_i \approx \mu_i - \frac{\sigma_i^2}{2}, i = 1, \cdots, M \tag{5.2}$$

where g_i, μ_i, and σ_i denote the geometric and arithmetic means and return volatility of individual assets, respectively. We can also apply Equation 5.1 to the FW portfolio as a whole to have

$$g_{FW} \approx \mu_{FW} - \frac{\sigma_{FW}^2}{2} \qquad (5.3)$$

This expression can be rearranged as

$$\mu_{FW} \approx g_{FW} + \frac{\sigma_{FW}^2}{2} \qquad (5.4)$$

Taking the weighted average of Equation 5.2 using portfolio weights $\left(w_1, \cdots, w_M \right)$ leads to

$$\sum_{i=1}^{M} w_i g_i \approx \sum_{i=1}^{M} w_i \mu_i - \frac{1}{2} \sum_{i=1}^{M} w_i \sigma_i^2 \qquad (5.5)$$

This is the same as

$$\bar{g} \approx \mu_{FW} - \frac{1}{2} \sum_{i=1}^{M} w_i \sigma_i^2 \qquad (5.6)$$

Substituting Equation 5.4 into Equation 5.6, we have

$$\bar{g} \approx g_{FW} + \frac{\sigma_{FW}^2}{2} - \frac{1}{2} \sum_{i=1}^{M} w_i \sigma_i^2 \qquad (5.7)$$

Rearranging the terms leads to

$$e_v = g_{FW} - \bar{g} \approx \frac{1}{2} \left(\sum_{i=1}^{M} w_i \sigma_i^2 - \sigma_{FW}^2 \right) \qquad (5.8)$$

The term on the right-hand side is the "diversification return." Since it is an approximation of the volatility effect, we shall label it as $\tilde{e}_v$. It has a matrix form

$$\tilde{e}_v = \frac{1}{2} \left(\mathbf{w}' \cdot \boldsymbol{\sigma}^2 - \mathbf{w}' \Sigma \mathbf{w} \right) \qquad (5.9)$$

The vector σ^2 is the vector of individual return variances.

- The result in Equation 5.8 is rather insightful. It approximates the difference of geometric returns, which span the entire sample period, by volatilities of the returns. It is appropriate to name it as volatility effect.

5.1.1 Two-Asset "Diversification Return"

When there are two assets in a portfolio, we have

$$
\begin{aligned}
\tilde{e}_v &= \frac{1}{2}\left[w_1\sigma_1^2 + w_2\sigma_2^2 - \left(w_1^2\sigma_1^2 + w_2^2\sigma_2^2 + 2w_1w_2\rho_{12}\sigma_1\sigma_2 \right) \right] \\
&= \frac{1}{2}\left(w_1w_2\sigma_1^2 + w_1w_2\sigma_2^2 - 2w_1w_2\rho_{12}\sigma_1\sigma_2 \right) \qquad (5.10) \\
&= \frac{1}{2}w_1w_2\left(\sigma_1^2 + \sigma_2^2 - 2\rho_{12}\sigma_1\sigma_2 \right)
\end{aligned}
$$

Of course, ρ_{12} is the correlation between the two assets. The term in the parentheses is the variance of the return difference between the two assets (see Equation 2.34). So, we have

$$
\tilde{e}_v = \frac{1}{2}w_1w_2\sigma_{1-2}^2 \qquad (5.11)
$$

The resulting Equation 5.11 is elegant in several ways. First, it is always non-negative for long-only portfolios, consistently with the analytic result from Chapter 4. It is zero only if one asset has zero weight or the return difference shows no variability, i.e. the return difference is constant through time. This is not the same analytic condition derived in Chapter 4 for zero volatility effect. But in a way, it is a close approximation of that condition (Problem 1).

Second, it is a product of weights and the variance of return difference. Therefore, the higher the variance, the higher the volatility effect. Intuitively, this makes sense, since the volatility of return difference determines the size of drift in weights away from the original weights. More drifts in weights present more opportunity for portfolio rebalancing.

Lastly, given the variance of return difference, the maximum "diversification return" is achieved when the portfolio is equally weighted in the two assets. This is because the maximum value of w_1w_2 is given by $w_1 = w_2 = 0.5$ when we have $w_1 + w_2 = 1$. When this is the case, $\tilde{e}_v = 0.125\sigma_{1-2}^2$.

TABLE 5.1   Volatility of Return Differences

	Stock Less Cash	Stock Less Bond	Commodity Less Stock	Commodity Less Bond
σ_{1-2}	17.63%	18.93%	31.51%	27.33%

TABLE 5.2   Volatility Effects and Diversification Returns of Three Portfolios

	50/50 Cash/Stock	50/50 Stock/Bond	50/50 Commodity/Stock	50/50 Commodity/Bond
e_v	0.39%	0.46%	1.14%	0.93%
$\tilde{e}_v$	0.39%	0.45%	1.24%	0.93%

Example 5.1: We use "diversification return" to estimate the volatility effect of four 50/50 asset allocation portfolios. They are 50/50 stock/cash, 50/50 stock/bond, 50/50 commodity/stock, and 50/50 commodity/bond portfolios. Table 5.1 displays the volatility of return differences between four pairs of assets, based on annual returns from 1970 to 2014. The stock/cash and stock/bond pairs have volatilities around 18%—they are of similar magnitude because stock's volatility dominates. The two pairs involving commodities have volatilities around 30%.

We use $\tilde{e}_v = 0.125\sigma_{1-2}^2$ to calculate "diversification return", which is shown in Table 5.2. The actual volatility effects are also presented. For the stock/cash, stock/bond, and commodity/bond portfolios, the estimation is very accurate. For commodity/stock portfolio, the estimate is 10 basis points (bps) higher than the actual volatility effect. As we discussed in Chapter 2, the arithmetic mean–geometric mean (AM-GM) approximation is less accurate when asset returns have high volatility. This might be the cause for the accuracy of the diversification effect. However, the estimation for the 50/50 commodity/bond portfolio is accurate despite relatively high volatility.

5.1.2 Pairwise Decomposition of "Diversification Return"

Equation 5.11 represents the diversification effect of rebalancing a pair of assets. For portfolios with more than two assets, portfolio rebalancing obviously involves all the assets in the portfolio. However, as we show later, the diversification return of any portfolio can be decomposed into a linear combination of diversification returns of all pairs.

Starting with the definition in Equation 5.8, we have

$$\tilde{e}_v = \frac{1}{2}\left(\sum_{i=1}^{M} w_i\sigma_i^2 - \sum_{i,j=1}^{M} w_iw_j\rho_{ij}\sigma_i\sigma_j\right) \qquad (5.12)$$

This equation can be rewritten as

$$2\tilde{e}_v = \sum_{i=1}^{M} w_i\sigma_i^2 - \sum_{i=1}^{M} w_i^2\sigma_i^2 - \sum_{i\neq j}^{M} w_iw_j\rho_{ij}\sigma_i\sigma_j$$

$$= \sum_{i=1}^{M} w_i\left(1-w_i\right)\sigma_i^2 - \sum_{i\neq j}^{M} w_iw_j\rho_{ij}\sigma_i\sigma_j \qquad (5.13)$$

$$= \sum_{i=1}^{M} w_i\sigma_i^2 \sum_{j\neq i}^{M} w_j - 2\sum_{i<j}^{M} w_iw_j\rho_{ij}\sigma_i\sigma_j$$

The first term can be rewritten as

$$\sum_{i=1}^{M} w_i\sigma_i^2 \sum_{j\neq i}^{M} w_j = \sum_{i<j}^{M} w_iw_j\left(\sigma_i^2 + \sigma_j^2\right) \qquad (5.14)$$

Substituting Equation 5.14 into Equation 5.13 and dividing the equation by 2, we have

$$\tilde{e}_v = \frac{1}{2}\sum_{i<j}^{M} w_iw_j\left(\sigma_i^2 + \sigma_j^2 - 2\rho_{ij}\sigma_i\sigma_j\right) = \frac{1}{2}\sum_{i<j}^{M} w_iw_j\sigma_{i-j}^2 \qquad (5.15)$$

Each term on the right-hand side is a diversification return of a pair of assets. Thus, the diversification return of a portfolio is the sum of diversification returns of all distinct pairs.

This result has several powerful consequences. First, it provides a direct proof that for long-only portfolios, diversification return is always nonnegative. This is, of course, consistent with the analytic result in Chapter 4.

We will discuss the case of long-short portfolios separately later. Second, the result is of a form that separates the weights and the variances of returns, and the dependence on the weights is quadratic, similarly to that of portfolio variance. As a matter of fact, we can write Equation 5.15 in a matrix form as follows:

$$\tilde{e}_v = \frac{1}{4} \mathbf{w}' \Sigma_v \mathbf{w} \tag{5.16}$$

The weight vector $\mathbf{w}$ consists of portfolio weights, and the matrix Σ_v is given by

$$\Sigma_v = \begin{pmatrix} 0 & \sigma_{1-2}^2 & \cdots & \sigma_{1-M}^2 \\ \sigma_{2-1}^2 & 0 & \cdots & \sigma_{2-M}^2 \\ \vdots & \vdots & \ddots & \vdots \\ \sigma_{M-1}^2 & \sigma_{M-2}^2 & \cdots & 0 \end{pmatrix} \tag{5.17}$$

Each term σ_{i-j}^2 is the variance of return difference between assets i and j. The fraction of 1/4 is to account for double counting of the number of pairs in the matrix.

5.1.3 Another Decomposition of "Diversification Return"

Diversification return of a portfolio can also be decomposed as a linear combination of diversification returns of M pairs, which is formed between M assets and the overall portfolio. We state the result in the following and then prove it. The decomposition is

$$\tilde{e}_v = \frac{1}{2} \left(\sum_{i=1}^{M} w_i \sigma_i^2 - \sigma_{FW}^2 \right) = \frac{1}{2} \sum_{i=1}^{M} w_i \sigma_{i-FW}^2 \tag{5.18}$$

The term σ_{i-FW}^2 is the variance of return differences between asset i and the FW portfolio. Since the weight of the whole portfolio is unity, each term in the summation can be thought of as diversification return between asset i and the FW portfolio.

We start with the identity

$$\sigma_{i-FW}^2 = \sigma_i^2 + \sigma_{FW}^2 - 2\text{cov}\left(r_i, r_{FW}\right) \tag{5.19}$$

Taking the weighted average leads to

$$
\begin{aligned}
\sum_{i=1}^{M} w_i \sigma_{i\text{-FW}}^2 &= \sum_{i=1}^{M} w_i \sigma_i^2 + \sum_{i=1}^{M} w_i \sigma_{\text{FW}}^2 - 2 \sum_{i=1}^{M} w_i \text{cov}\left(r_i, r_{\text{FW}}\right) \\
&= \sum_{i=1}^{M} w_i \sigma_i^2 + \sigma_{\text{FW}}^2 - 2\,\text{cov}\left(\sum_{i=1}^{M} w_i r_i, r_{\text{FW}}\right) \\
&= \sum_{i=1}^{M} w_i \sigma_i^2 + \sigma_{\text{FW}}^2 - 2 \sigma_{\text{FW}}^2 \\
&= \sum_{i=1}^{M} w_i \sigma_i^2 - \sigma_{\text{FW}}^2
\end{aligned}
\tag{5.20}
$$

This completes the proof of Equation 5.18. The interpretation of this decomposition seems to be that portfolio rebalancing can be viewed as rebalancing between individual assets and the overall portfolio. In a long-only setting, when an asset outperforms (underperforms) the portfolio, one would sell (buy) the asset and buy (sell) the portfolio.

- These decompositions seem to be limited to diversification return; it does not extend to the original volatility effect. In other words, volatility effect of a portfolio is not simply a sum of volatility effects of pairs or a sum of volatility effect between individual assets and the overall portfolio.

5.2 MAXIMIZING "DIVERSIFICATION RETURN"

In the two-asset case, the diversification return is at its maximum when the two assets are equally weighted in the portfolio. We can pose the same question when portfolios have more than two assets. Diversification return, by itself, is not alpha. However, it is one part of it, and the solution of the maximization problem will reveal insights about the solution of maximizing alpha later in the book.

The maximization problem can be worked out in two different settings. The first is the original derivation of diversification effect, and the second is the pairwise decomposition of diversification return.

With diversification defined in Equation 5.9, we maximize $\mathbf{w}' \cdot \sigma^2 - \mathbf{w}' \Sigma \mathbf{w}$ subject to the constraint $\mathbf{w}' \cdot \mathbf{i} = 1$. If we do not impose the long-only condition, the solution is given by

$$\mathbf{w}^* = \frac{1}{2}\left(\Sigma^{-1}\sigma^2 - \frac{\mathbf{i}'\Sigma^{-1}\sigma^2 - 2}{\mathbf{i}'\Sigma^{-1}\mathbf{i}}\Sigma^{-1}\mathbf{i}\right) \tag{5.21}$$

The solution is no doubt correct, but it lacks intuition. The pairwise decomposition of diversification return is preferred in this regard. Before we derive the general solution of maximal diversification return, we present a detailed analysis for portfolios with three-assets, in which case

$$2\tilde{e}_v = w_1 w_2 \sigma_{1-2}^2 + w_2 w_3 \sigma_{2-3}^2 + w_3 w_1 \sigma_{3-1}^2 \tag{5.22}$$

The portfolio weights that maximize it are given by

$$w_1 = \frac{\sigma_{1-2}^2 + \sigma_{3-1}^2 - \sigma_{2-3}^2}{\sigma_{1-2}^2 \sigma_{3-1}^2} l$$

$$w_2 = \frac{\sigma_{2-3}^2 + \sigma_{1-2}^2 - \sigma_{3-1}^2}{\sigma_{2-3}^2 \sigma_{1-2}^2} l \tag{5.23}$$

$$w_3 = \frac{\sigma_{3-1}^2 + \sigma_{2-3}^2 - \sigma_{1-2}^2}{\sigma_{3-1}^2 \sigma_{2-3}^2} l$$

where the parameter l is a scaling factor such that $w_1 + w_2 + w_3 = 1$. We have

$$l = \left(\frac{\sigma_{1-2}^2 + \sigma_{3-1}^2 - \sigma_{2-3}^2}{\sigma_{1-2}^2 \sigma_{3-1}^2} + \frac{\sigma_{2-3}^2 + \sigma_{1-2}^2 - \sigma_{3-1}^2}{\sigma_{2-3}^2 \sigma_{1-2}^2} + \frac{\sigma_{3-1}^2 + \sigma_{2-3}^2 - \sigma_{1-2}^2}{\sigma_{3-1}^2 \sigma_{2-3}^2}\right)^{-1} \tag{5.24}$$

Since the total diversification return consists of a weighted average of three pairwise variances of return differences, the optimal weights would tend to maximize the coefficients for large pairwise variances. It is evident from Equation 5.23 that when three pairwise variances are identical, the portfolio weights will be equal for the three assets. When they are not equal, the weight will be higher for the assets that are associated with higher pairwise variances. For instance, asset 1 is involved in two pairs, one with Asset 2 and the other with Asset 3. Hence, its weight in Equation 5.23 is positively influenced by the two pairwise variances σ_{1-2}^2 and σ_{3-1}^2. On the other hand, it is not involved in the remaining pair between Assets 2 and 3. Its weight is then negatively influenced by the variance for that pair, σ_{2-3}^2. The weight is positive when $\sigma_{1-2}^2 + \sigma_{3-1}^2 - \sigma_{2-3}^2$ is greater than zero and negative when $\sigma_{1-2}^2 + \sigma_{3-1}^2 - \sigma_{2-3}^2$ is lower than zero.

This intuition carries over to the general cases with M assets. If we combine the matrix representation of the diversification return and the budget constraint $\mathbf{w}' \cdot \mathbf{i} = 1$, the optimal weight vector is

$$\mathbf{w}^* = \frac{\Sigma_v^{-1} \mathbf{i}}{\mathbf{i}' \Sigma_v^{-1} \mathbf{i}} \tag{5.25}$$

This solution takes the same form as the portfolio weights of a minimal variance portfolio except that the matrix involved is the matrix of pairwise variances instead of a traditional covariance matrix. However, at first glance, it seems peculiar that Equation 5.25 should *maximize* the diversification return. For one thing, unlike a return covariance matrix, the matrix Σ_v is not positive definite.

We resolve these issues by noting two aspects of the optimization problem. First, if we reduce the dimension of the problem by 1 with the use of budget constraint $\mathbf{w}' \cdot \mathbf{i} = 1$, then the problem would correspond to a traditional quadratic programming problem with a positive definite matrix. In a two-asset case, the objective function changes from $w_1 w_2$ to $w_1 (1 - w_1)$, which is well defined and has a maximum at $w_1 = 0.5$. Second, it can be seen that Equation 5.25 is the solution of the reduced optimization problem.

Example 5.2: For a three-asset portfolio with stocks, bonds, and commodities, we have the pairwise variances exhibited in Table 5.1. The corresponding matrix Σ_v would be

$$\Sigma_v = \begin{pmatrix} 0 & (18.93\%)^2 & (31.51\%)^2 \\ (18.93\%)^2 & 0 & (27.33\%)^2 \\ (31.51\%)^2 & (27.33\%)^2 & 0 \end{pmatrix} \tag{5.26}$$

We have designated stocks, bonds, and commodities as Asset 1, 2, and 3, respectively. Using Equation 5.25 gives

$$\mathbf{w}^* = \begin{pmatrix} 42.7\% \\ 10.5\% \\ 46.8\% \end{pmatrix} \tag{5.27}$$

The maximum diversification return is achieved with 42.7%, 10.5%, and 46.8% in stocks, bonds, and commodities, respectively. According to either Equation 5.22 or Equation 5.16, the diversification return is 1.26%.

- The optimal weight for bonds is quite low since its volatility is quite low, resulting in low pairwise variances. Nevertheless, the bond weight is positive. If we choose not to invest in bonds, then the optimal weights should be 50% in stocks and 50% in commodities. This combination has a diversification return of 1.24%, which is only 2 bps less.

5.3 DIVERSIFICATION RETURNS OF LONG-SHORT PORTFOLIOS

As we saw in Chapter 3, the dynamics of rebalancing long-short portfolios is different from that of long-only portfolios. In a long-only portfolio, rebalancing is achieved by selling winners and buying losers. As a result, rebalancing takes advantage of price changes caused by return volatility and generated positive volatility effect. The diversification return is positive as well. In contrast, in a long-short portfolio, rebalancing is more complicated. Depending on the makeup of a portfolio, two actions— selling winners/buying losers as well as buying winners/selling losers— could be involved. Naturally, the action of buying winners/selling losers would give rise to a negative contribution to volatility effect. The task is to combine both positive and negative contributions together.

5.3.1 Two-Asset Long-Short Portfolios

The diversification return of two-asset long-short portfolios is always negative. This is evident from Equation 5.11, i.e. $\tilde{e}_v = (1/2)w_1 w_2 \sigma_{1-2}^2$. With a long-short portfolio, one of the weights is positive and the other is negative. Therefore, the product is negative.

This formula is intuitive in several respects. First, the diversification return is zero if $w_1 = 0$ or $w_1 = 1$, i.e. when the portfolio is invested 100% in either asset. There is no rebalance and hence, no diversification return. Second, the diversification is positive when $0 < w_1 < 1$, in which case the portfolio is long-only and unlevered. However, the diversification becomes negative when $w_1 < 0$ or $w_1 > 1$; in other words, when the portfolio is long-short.

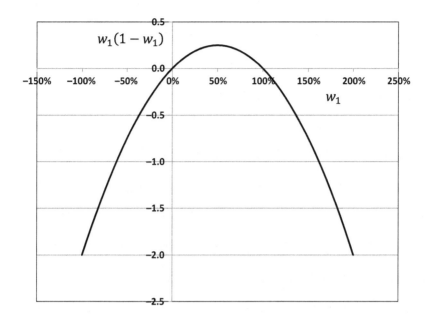

FIGURE 5.1 The quadratic function $w_1(1-w_1)$ that shows the relationship between the weight and the diversification return.

Figure 5.1 plots the quadratic function of the weight. It not only demonstrates the intuition discussed above but also reveals the fact that the maximum diversification return is achieved when the portfolio is 50/50.

- We have proved in Chapter 4 that the volatility effect of two-asset long-short portfolios is always negative.

When the short asset is the risk-free asset, then the long-short portfolio can be considered as a levered portfolio in the risky asset. We designate Asset 1 as the risk-free asset and denote the weight of Asset 2 as L. Then, the diversification return is written as

$$\tilde{e}_v = \frac{1}{2}\left(L - L^2\right)\sigma^2 \tag{5.28}$$

We can interpret L as the leverage and σ as the volatility of the risky asset.

5.3.2 Inverse and Leveraged Exchange-Traded Funds

An ETF is an exchange-traded fund. Traditional mutual funds can only be bought and sold once a day at a closing price that reflects the net asset value of the fund at the end of trading. An ETF is an investment-like mutual fund, but it trades like a stock throughout the day on a stock exchange. Most ETFs are based on investment indices in stocks, bonds, and commodities. For example, the ETF SPY is based on the S&P 500 index, and over time, the performance of "SPY" should follow closely the return of the S&P 500 index, net of management fee of the ETF, transaction costs, and possible tracking error between the ETF and the index. Compared with traditional index mutual funds, ETFs offer the convenience of trading like stocks.

Over the years, the popularity of ETFs has grown tremendously. Not only has the number of assets under management by ETF products increased exponentially, but also the new types of ETFs offered to investors. These new types include inverse ETFs and leveraged ETFs. An inverse ETF represents an investment shorting an index. For example, "SH" is an ETF that is short the S&P 500 index and aims to deliver returns that are opposite to the S&P 500 index on a daily basis. On the other hand, a leveraged ETF is designed to deliver returns that are multiples of a traditional index. For example, "SSO" is an ETF designed to deliver returns that are twice the return of the S&P 500 index on a daily basis. In addition, a leveraged ETF could also be short the index, with a negative multiple of two or even three. When the multiple is positive, the ETF is referred to as ultralong, and when the multiple is negative, the ETF is labelled as ultra-short.

Inverse and leveraged ETFs are different from other ETFs in terms of portfolio rebalancing. Since many underlying indices for ETFs are capitalization-weighted indices, a long-only ETF is essentially a buy-and-hold portfolio, with no rebalancing. For inverse and leveraged ETFs, the goal is to deliver returns that are a multiple of the index returns on a daily basis, and the underlying portfolios need to be rebalanced every day. This daily rebalancing generates a negative diversification return, given by Equation 5.28.

To determine the size of the potential return slippages or negative diversification returns, we provide numerical examples by assuming that the risk index has an annual volatility of 20% (the likes of the S&P 500 index). Then, the daily volatility is about 1.3% ($20\%/\sqrt{250}$), assuming that daily returns are independent. Using Equation 5.28, we obtain the daily return slippage in Table 5.3 for the following five ETFs: –3X, –2X, –1X, 2X, and 3X, with the letter X indicating multiples. The next row displays the annualized slippage. The annual slippage of –3X ETF is a whopping –21.3%, while that of 2X ETF is –11.3%. These return slippages

TABLE 5.3 Daily and Annualized Diversification Returns (Slippages) of Inverse and Leveraged ETFs of an Index Whose Annual Volatility is 20%

	−3X	−2X	−1X	2X	3X
Daily	−0.10%	−0.05%	−0.02%	−0.02%	−0.05%
Annual	−21.3%	−11.3%	−3.9%	−3.9%	−11.3%

create significant return hurdles for these leveraged and inverse (or both) ETFs. For example, when the index annual return is −5%, a naïve investor might expect the super-bearish −3X ETF to yield +15%. But in reality, the likely outcome is −6.3% (=15%−21.3%). An even more puzzling case could happen: that in a year when the index is down, −3X ETF underperforms −2X ETF.

This example of levered portfolios of a single risky asset seems to indicate that shorting and leverage leads to negative diversification returns. Is this also the case with multiple risky assets portfolios? The answer is not necessarily. The reason is that those portfolios have the benefit of diversification among multiple assets, which is lacking in the present case.

5.3.3 Leveraged "Long-Only" Portfolios

A special case of long-short portfolios is one in which the only short position is in the risk-free assets, and all others are long positions in risky assets. A practical example of such a portfolio is risk parity portfolios, which invest in asset classes such as stocks, bonds, and commodities with a balanced risk allocation and use portfolio leverage to achieve targeted portfolio risks and returns.

In such a leveraged "long-only" portfolio, there are two opposing forces affecting its diversification return. On one side, since the portfolio remains long-only, the positive diversification effect among the assets remains. On the other side is the leverage itself, of the whole portfolio, which tends to generate negative diversification return, as we have shown earlier. We analyze the two effects and their contributions.

Assuming that the short asset is Asset 1 and the long assets are from Asset 2 to M, the leverage ratio of the portfolio is the sum of all risky assets:

$$\sum_{i=2}^{M} w_i = L \tag{5.29}$$

The weight in the risk-free asset is $w_1 = 1 - L$. We now rescale the weights of the risky assets by the leverage and denote them as $\tilde{w}_i = w_i/L$. These rescaled weights sum up to 1:

$$\sum_{i=2}^{M} \tilde{w}_i = 1 \tag{5.30}$$

Assuming that the volatility of the risk-free asset is zero, the diversification return of this portfolio is

$$\tilde{e}_v = \frac{1}{2}\left(\sum_{i=2}^{M} w_i\sigma_i^2 - \sum_{i,j=2}^{M} w_i w_j \rho_{ij}\sigma_i\sigma_j \right) \tag{5.31}$$

We now express the weights in terms of scaled weights and the leverage ratio. We have

$$\tilde{e}_v = \frac{1}{2}\left(L\sum_{i=2}^{M} \tilde{w}_i\sigma_i^2 - L^2\sum_{i,j=2}^{M} \tilde{w}_i \tilde{w}_j \rho_{ij}\sigma_i\sigma_j \right) \tag{5.32}$$

We rewrite it as

$$\tilde{e}_v = \frac{L}{2}\left(\sum_{i=2}^{M} \tilde{w}_i\sigma_i^2 - \sum_{i,j=2}^{M} \tilde{w}_i \tilde{w}_j \rho_{ij}\sigma_i\sigma_j \right) + \frac{1}{2}\left(L - L^2\right)\sum_{i,j=2}^{M} \tilde{w}_i \tilde{w}_j \rho_{ij}\sigma_i\sigma_j \tag{5.33}$$

We now recognize that the term inside the parentheses of the first term is the diversification return of the long-only portfolio with the rescaled weights. Moreover, the double summation in the second term is the variance of that rescaled portfolio. Therefore, Equation 5.33 can be expressed as

$$\tilde{e}_v = L\tilde{e}_{v,s} + \frac{1}{2}\left(L - L^2\right)\sigma_s^2 \tag{5.34}$$

Equation 5.34 provides an intuitive interpretation of two sources of diversification return of the levered portfolio relative to the diversification return of the scaled unlevered portfolio. The first source is positive; it is the leverage ratio times the diversification return of the rescaled portfolio $\tilde{e}_{v,s}$. The second term is of the same form as Equation 5.28; it is the negative diversification return of levering a single asset—in this case, the rescaled portfolio. It is a function of the leverage ratio and the volatility of the scaled unlevered portfolio σ_s.

Consider three cases of leverage ratio. First, when $L = 1$, there is no leverage, and the second term vanishes. The diversification return is that of a long-only portfolio. Second, when $L > 1$, the first term increases the positive diversification return of the scaled long-only portfolio, but the second term is negative. The net result could be either positive or negative, depending on the size of leverage, the volatility, and the diversification return of the rescaled portfolio. Lastly, the result is equally applicable when $L < 1$ for a portfolio that is not fully invested. In this case, both terms are positive. The portfolio is essentially long-only and unlevered with M risky assets and one risk-free asset, and the diversification return is always positive.

Example 5.3: We illustrate this decomposition with an example of a risk parity portfolio consisting of stocks, bonds, and commodities. For illustration purposes, we use in-sample volatilities of asset classes to compute the notional weights. The volatilities are listed in Table 5.4. The risk parity weights are inversely proportional to volatilities, and notional exposures are calibrated such that the portfolio has a historical volatility of 10%. It has a leverage ratio of about 143%. Dividing the risk parity portfolio weights by the leverage yields the scaled weights of 57.48% in bonds, 23.99% in stocks, and 18.53% in commodities. This scaled unlevered portfolio has a volatility of about 7%.

Using Equation 5.33, we calculate that the diversification return of the scaled portfolio is 0.87%. Consequently, the diversification return of the risk parity portfolio is

$$\tilde{c}_v - 143\% \cdot 0.87\% + \frac{1}{2}\left[143\% - \left(143\%\right)^2\right]\left(7\%\right)^2 \tag{5.35}$$

$$= 1.24\% - 0.15\% = 1.09\%$$

TABLE 5.4 Asset Volatility, Risk Parity Portfolio Weights, and Scaled Portfolio Weights

	Bonds	**Stocks**	**Commodities**	**Portfolio Risk**	**L**
Volatility	7.69%	17.43%	24.93%		
Risk parity weights	82.16%	34.29%	26.48%	10%	143%
Scaled weights	57.48%	23.99%	18.53%	7%	100%

The total diversification return of the risk parity portfolio is 1.09%, of which 1.24% is the positive contribution from the leveraged long-only portfolio and −0.15% is the negative contribution due to leverage.

Equation 5.34 presents the diversification return as a quadratic function of the leverage ratio once the underlying portfolio is fixed. Since the coefficient for the square term is negative, the diversification return will become negative when the magnitude of leverage becomes large enough. It is zero when $L = 0$ and when the leverage reaches

$$L = 1 + \frac{2\tilde{e}_{v,s}}{\sigma_s^2} \tag{5.36}$$

In Example 5.3, this value is 355%, beyond which the diversification return would be negative.

The diversification return is at its maximum when the leverage is

$$L = \frac{1}{2} + \frac{\tilde{e}_{v,s}}{\sigma_s^2} \tag{5.37}$$

This value is 228% in the example.

Another perspective of diversification of long-short portfolios is through pairwise diversifications. Inspection of Equation 5.15 shows that pairs with weights of the same sign, either both long or both short, contribute positively to the overall diversification return. On the other hand, pairs with different signs, i.e. long and short, contribute negatively to the overall diversification return.

PROBLEMS

5.1. The condition for zero volatility effect is $1 + \mathbf{r}_1 = k(1 + \mathbf{r}_2)$, with $\mathbf{r}_1$ and $\mathbf{r}_2$ denoting the vectors of returns for the two assets and k being a positive constant. Show that this condition can be approximated by $\sigma_{1-2}^2 = 0$ when the returns are small.

5.2. (a) Derive the weight vector in Equation 5.21 that maximizes diversification return. (b) Prove that when there are two assets, Equation 5.21 reduces to $\mathbf{w}^* = \left(1/2, 1/2\right)'$, i.e. the weight of an equally weighted portfolio.

5.3. Derive the optimal weights in Equation 5.23.

5.4. Prove that the scaling parameter given by Equation 5.24 is positive.

5.5. Derive the results in Example 5.3.

CHAPTER 6

Analysis of Return Effect

W E STUDY THE RETURN effect of portfolio rebalancing in this chapter. The return effect is the difference between the return of a buy-and-hold (BH) portfolio and the weighted average of geometric means of individual assets in the portfolio. We have proved in Chapter 4 that the return effect is always non-negative for long-only portfolios. We shall first derive approximations of the return effects for long-only portfolios. As we shall see, in the case of return effect, long-short portfolios require special treatment because of the negative weights involved, for which standard results such as Jensen's inequality and Taylor expansion are no longer valid.

The return effect is greatly influenced by serial cross-sectional correlations of asset returns. It is commonly acknowledged that BH portfolios benefit from trending or momentum, while fixed-weight (FW) portfolios benefit from reversal with portfolio rebalancing. This influence of serial correlations on the two types of portfolios is through return effect. In this chapter, we quantify the impact of cross-sectional correlation on the return effect.

6.1 RETURN EFFECT OF LONG-ONLY PORTFOLIOS

From Equation 4.2, we have

$$\left(1 + g_{\mathrm{BH}}\right)^N = w_1 \left(1 + g_1\right)^N + \cdots + w_M \left(1 + g_M\right)^N \tag{6.1}$$

To relate the right-hand side to the weighted average of geometric returns $\bar{g}$, we approximate each term $(1+g_i)^N$ by $(1+\bar{g})^N$ using a second-order Taylor expansion,

$$(1+g_i)^N = (1+\bar{g})^N \left(1+\frac{g_i-\bar{g}}{1+\bar{g}}\right)^N$$

$$\approx (1+\bar{g})^N \left[1+N\frac{g_i-\bar{g}}{1+\bar{g}}+\frac{N(N-1)}{2}\left(\frac{g_i-\bar{g}}{1+\bar{g}}\right)^2\right]$$

(6.2)

The weighted sum of Equation 6.2 gives rise to

$$(1+g_{\mathrm{BH}})^N = \sum_{i=1}^{M} w_i (1+g_i)^N \approx (1+\bar{g})^N \left[1+\frac{N(N-1)}{2(1+\bar{g})^2}\sum_{i=1}^{M} w_i (g_i-\bar{g})^2\right] \quad (6.3)$$

The linear term drops out by the definition of $\bar{g}$. We define the weighted variance of the geometric returns across the assets as

$$\mathrm{var}(g) = \sum_{i=1}^{M} w_i (g_i-\bar{g})^2$$

(6.4)

Some caution is necessary regarding this definition of generalized variance. When the portfolio is equally weighted with $w_i = 1/M$, the average $\bar{g}$ and the variance $\mathrm{var}(g)$ are the simple arithmetic average and variance. When the weights are positive but not equal, $\mathrm{var}(g)$ can be interpreted as the variance of returns with discrete probabilities given by $w_i, i = 1,...,M$. However, when the weights are from a long-short portfolio, $\mathrm{var}(g)$ is not a conventional variance.

Re-approximating the bracket term on the right hand side of Equation 6.3 as a power, we have

$$(1+g_{\mathrm{BH}})^N \approx (1+\bar{g})^N \left[1+\frac{(N-1)}{2(1+\bar{g})^2}\mathrm{var}(g)\right]^N$$

$$= \left[1+\bar{g}+\frac{(N-1)}{2(1+\bar{g})}\mathrm{var}(g)\right]^N$$

(6.5)

Taking the root of both sides yields an approximation of the return effect. We have

$$e_r = g_{BH} - \overline{g} \approx \tilde{e}_r = \frac{(N-1)}{2(1+\overline{g})} \operatorname{var}(g) \tag{6.6}$$

Equation 6.6 is the main result for the return effect of a long-only portfolio. We make several remarks about the results.

- First, the weighted variance of geometric returns is positive when all weights are non-negative. Therefore, Equation 6.6 gives rise to a positive return effect for long-only portfolios, consistently with the inequality $g_{BH} \geq \overline{g}$.

- Second, we consider the validity of the approximation used to obtain this result. For the sake of Taylor expansion in Equation 6.2, the differences $g_i - \overline{g}$ should be small and somewhat comparable. If one of the geometric returns, say g_k, is dominant, and w_k is moderately positive, then the sum $\sum_{i=1}^{M} w_i (1 + g_i)^N$ will be dominated by $w_k (1 + g_k)^N$ when N becomes large. In other words, if one particular asset has a much higher return than the rest, the return of a BH portfolio, g_{BH}, will largely depend on that asset, rendering Equation 6.6 invalid. Imagine we have a BH portfolio consisting of cash and equity. Over the long run, the equity return and its initial allocation are all that matters. In this case, Equation 6.6 will not be a good approximation.

- Another condition is that the weighted variance $\operatorname{var}(g)$ has to be reasonably small for Equation 6.5 to hold. This tends not to be an issue for long-only portfolios. When weights are all non-negative, the weighted variance is a true variance associated with a set of discrete probabilities. However, it is a potential problem for long-short portfolios where some weights are negative. This is because with some negative weights, $\operatorname{var}(g)$ is no longer a true variance, and the weighted average $\overline{g}$ is not a true average either. For example, $\overline{g}$ could lie outside the range of individual returns. In that case, $\sum_{i=1}^{M} w_i (g_i - \overline{g})^2$ could be rather large, since some g_i could be quite far away from $\overline{g}$. Therefore, for long-short portfolios, Equation 6.6 could be less satisfactory. We shall derive an alternative version that provides a more accurate approximation later in the chapter.

6.1.1 Two-Asset Return Effect

When there are just two assets, the weighted variance is

$$\mathrm{var}\left(g\right) = w_1\left(g_1 - w_1 g_1 - w_2 g_2\right)^2 + w_2\left(g_2 - w_1 g_1 - w_2 g_2\right)^2$$
$$= w_1 w_2 \left(g_1 - g_2\right)^2 \tag{6.7}$$

The variance is a product of weights of the two assets and the square of the difference of two returns. Therefore, it is positive if both assets have positive weights for a long-only portfolio. However, it is negative when the weights are of opposite signs for a long-short portfolio. Hence, the approximation given by Equations 6.6 and 6.7 is at least consistent with the fact that the return effect of a two-asset long-short portfolio is negative, which was established in Chapter 4. But there is reason to believe that this approximation might not be accurate because of negative portfolio weights. We shall derive an alternative approximation of return effect for long-short portfolios later in the chapter.

> **Example 6.1:** We estimate return effect of four 50/50 asset alloca-
> tion portfolios. They are 50/50 stock/cash, 50/50 stock/bond, 50/50
> commodity/stock, and 50/50 commodity/bond portfolios. These
> are the same portfolios examined in Example 5.1 for their volatility
> effects.
>
> Table 6.1 lists the geometric return of the four asset classes from
> 1970 to 2014, from which we can calculate the approximation of the
> return effects of the four portfolios using Equations 6.6 and 6.7. We
> compare these results with the actual return effects from the return
> of BH portfolios and the weighted average of geometric returns. The
> results are listed in Table 6.2.
>
> We make several comments about the results. First, the magnitude
> of return effect is closely related to the difference of the geometric
> returns of the assets involved. According to Table 6.1, the difference
> between returns of bond and commodity is rather small. As a result,
> the return effect of the 50/50 commodity/bond portfolio is just 0.01%.
> On the other hand, there is quite a separation between the returns of

TABLE 6.1   Geometric Asset Class Returns from 1970 to 2014

	Cash	Bond	Stock	Commodity
g	5.16%	7.58%	10.48%	8.03%

TABLE 6.2   Return Effects and Approximations of Four Portfolios

	50/50 Cash/Stock	50/50 Stock/Bond	50/50 Commodity/Stock	50/50 Commodity/Bond
e_r	1.22%	0.40%	0.29%	0.01%
$\tilde{e}_r$	1.44%	0.42%	0.30%	0.01%

cash and stock. Consequently, the return effect of 50/50 cash/stock portfolio is rather high, at 1.22%. Second, the approximation of the return effect is excellent when the return effect itself is small. This is because the error in the approximation is small when the returns of assets are of similar magnitude, which also leads to low return effect. The exception is the 50/50 cash/stock portfolio. The stock return is significantly higher than cash return, creating large return effect as well as larger error in the approximation.

Example 6.2: We now consider a two-asset long-short portfolio and examine both return and volatility effects and their approximations. The long-short portfolio has 150% long in stocks and −50% in cash, which can be implemented by investing in equity index futures and treasury bills. The average geometric return is $\bar{g} = 1.5 \cdot 10.48\% - 0.5 \cdot 5.16\% = 13.14\%$. The return of the FW portfolio is $g_{FW} = 11.81\%$, and the return of the BH portfolio is $g_{BH} = 11.39\%$. These lead to volatility and return effect of −1.33% and −1.75%, respectively, as shown in Table 6.3.

As we have proved in Chapter 4, both effects are indeed negative for this long-short portfolio. When we calculate the approximation of the volatility effect using Equation 5.10 or Equation 5.11, we have $\tilde{e}_v = -1.17\%$. This approximation is reasonably accurate, given the fact that the volatility of this leveraged equity portfolio is rather high. However, when we calculate the approximation of the return effect using Equations 6.6 and 6.7, we obtain $\tilde{e}_r = -4.13\%$. While the error for the volatility effect is moderate, the error for the return effect is significantly large, indicating a potential fundamental error in the result. We shall address this issue later in the chapter.

TABLE 6.3   Volatility and Return Effects of a Long-Short Portfolio with Long 150% in Stocks and Short 50% in Cash

	Volatility Effect	Return Effect
Actual	−1.33%	−1.75%
Approximation	−1.17%	−4.13%

6.1.2 Pairwise Decomposition of Return Effect

The cross-sectional variance of geometric returns can be similarly decomposed into pairwise variances of all possible pairs. We have proved this result in Chapter 2. Equation 2.11 leads to

$$\operatorname{var}(g) = \sum_{i=1}^{M} w_i (g_i - \bar{g})^2 = \frac{1}{2} \sum_{i,j=1}^{M} w_i w_j (g_i - g_j)^2 \tag{6.8}$$

We recognize that each term inside the summation is variance between a pair of assets. When we combine Equations 6.8 and 6.6, the approximation of the return effect is

$$\tilde{e}_r = \frac{(N-1)}{4(1+\bar{g})} \sum_{i,j=1}^{M} w_i w_j (g_i - g_j)^2 = \frac{(N-1)}{2(1+\bar{g})} \sum_{i<j}^{M} w_i w_j (g_i - g_j)^2 \tag{6.9}$$

The second summation is over all distinct pairs. In addition, when $\bar{g}$ is small, we can drop the term $1 + \bar{g}$ from the denominator.

Similarly to the approximation of the volatility effect, there is a matrix for Equation 6.9. Define

$$\Sigma_g = \begin{pmatrix} 0 & (g_1-g_2)^2 & \cdots & (g_1-g_M)^2 \\ (g_2-g_1)^2 & 0 & \cdots & (g_2-g_M)^2 \\ \vdots & \vdots & \ddots & \vdots \\ (g_M-g_1)^2 & (g_M-g_2)^2 & \cdots & 0 \end{pmatrix} \tag{6.10}$$

Then

$$\tilde{e}_r = \frac{(N-1)}{4(1+\bar{g})} \mathbf{w}' \Sigma_g \mathbf{w} \tag{6.11}$$

We make several comments about the matrix representation.

- Equation 6.11, similarly to the matrix form of diversification return in Chapter 5 (Equation 5.16), is also of quadratic form in the weight vector. This would make the analysis of rebalancing alpha analytically tractable in the next chapter.

- The matrix Σ_g, while symmetric, is singular when the number of assets exceeds three.

- For a two-asset long-only portfolio, the maximum return effect is achieved with an equal-weighted portfolio. Since every term in Equation 6.11 is positive and separable in weights and returns, the maximum return effect is given by an equal-weighted portfolio of the two assets that have maximum difference in their geometric returns.

6.2 THE IMPACT OF CROSS-SECTIONAL SERIAL CORRELATIONS ON RETURN EFFECT

According to the results from the previous section, the return effect is approximately proportional to the generalized variance of geometric returns. The geometric return, by its very nature, is a cumulative return over the entire period. Returns from all periods contribute to geometric returns. Therefore, it is not hard to see that statistically, serial correlations will play a role in determining the overall geometric return. For a single asset, positive (negative) serial autocorrelation will result in higher (lower) geometric return. For a group of assets, positive (negative) cross-sectional serial correlations will result in higher (lower) cross-sectional dispersion of geometric returns. Positive serial correlations are often referred to as *momentum*, while negative ones are referred to as *reversal*.

For long-only portfolios, it is apparent that return reversal is beneficial to FW portfolios, and return momentum is beneficial to BH portfolios. While this intuition is generally true, we now employ Equation 6.6 to analyze the impact of serial correlations on the variance of the geometric returns.

First, we note that geometric return can be approximated by arithmetic return minus half of the variance. With $g \approx \mu - \sigma^2/2$, we have

$$\text{var}(g) \approx \text{var}(\mu) + \frac{1}{4}\text{var}(\sigma^2) - \text{cov}(\mu, \sigma^2) \qquad (6.12)$$

In cases where all assets have similar return volatilities, the variability of arithmetic returns of different assets becomes the dominant factor. We can then focus on the impact of serial correlations on the variances of arithmetic returns. Note that the last term in Equation 6.12 also involves the variance of arithmetic returns, and its impact on the variance of geometric return depends on the sign of correlation between μ and σ^2. If the correlation is negative, then a lower $\text{var}(\mu)$ will certainly lead to a lower $\text{var}(g)$.

By definition, arithmetic returns are simply arithmetic averages of returns over N periods. We can write

$$\begin{pmatrix} \mu_1 \\ \vdots \\ \mu_M \end{pmatrix} = \frac{1}{N} \begin{pmatrix} r_{11} + r_{12} + \cdots + r_{1N} \\ \vdots \\ r_{M1} + r_{M2} + \cdots + r_{MN} \end{pmatrix} \tag{6.13}$$

It is more convenient to express the equation in a vector form:

$$\mu = \frac{1}{N}(\mathbf{r}_1 + \mathbf{r}_2 + \cdots + \mathbf{r}_N) \tag{6.14}$$

Each vector denotes returns across M assets for a particular period. Taking generalized variance of both sides with portfolio weights,* we have

$$\text{var}(\mu) = \frac{1}{N^2}\text{var}(\mathbf{r}_1 + \mathbf{r}_2 + \cdots + \mathbf{r}_N) \tag{6.15}$$

The variance of a sum of vectors can be written as a sum of variances of individual vectors and their covariances, which in turn are products of cross-sectional serial correlations and dispersions of return vectors. We have

$$\text{var}(\mu) = \frac{1}{N^2}\left[\sum_{j=1}^{N} \text{var}(\mathbf{r}_j) + 2\sum_{j<k}^{N} \rho(\mathbf{r}_j, \mathbf{r}_k)\sigma(\mathbf{r}_j)\sigma(\mathbf{r}_k) \right] \tag{6.16}$$

We make several remarks.

- Equation 6.16 is analogous to the formula for calculating the variance of multi-period returns of one asset from variances of single-period returns and serial autocorrelations of the asset—a process often referred to as *annualizing* (see Section 2.5). The slight difference is that here we are interested in the variance of average return, not the variance of cumulative return.

---

* When we have an equally weighted portfolio, the variance will be a conventional one.

- It is apparent from Equation 6.16 that negative cross-sectional serial correlations would lead to lower variance of arithmetic means and thus smaller return effect from portfolio rebalancing, leading to larger rebalancing alpha in favor of FW portfolios.

When the cross-sectional serial correlations are all zero, we have

$$\mathrm{var}(\mu) = \frac{1}{N^2} \sum_{j=1}^{N} \mathrm{var}(\mathbf{r}_j) = \frac{1}{N} \overline{\mathrm{var}(\mathbf{r})} \qquad (6.17)$$

In other words, the variance of average return is the average of cross-sectional variances divided by N. One concludes that the variance of arithmetic means (not cumulative return) over the entire period is much smaller than the average of single-period variance.*

We note from Equation 6.16 that the cross-sectional serial correlations cover all lags from 1 to $(N-1)$. It is often the case that for some lags, the correlations are positive (cross-sectional momentum), while for other lags, the correlations are negative (cross-sectional reversal). It is the aggregated impact of correlations of all lags that determines their impact on the variance of arithmetic returns. For example, many asset returns exhibit short-term momentum. It would be incorrect to conclude on this basis alone that rebalancing alpha will be negative. The reason is that these asset returns might also have long-term reversal, which portfolio rebalancing will capture over the long term. We must weigh their effects accordingly.

Some theoretical simplification of Equation 6.16 would make these points clearer. Assuming that the variances of return vectors are the same for N periods, and the correlations between return vectors of different periods are stationary, i.e. they only depend on the lag and not on specific time periods, Equation 6.16 simplifies to

$$\mathrm{var}(\mu) = \frac{\mathrm{var}(\mathbf{r})}{N} \left[1 + \frac{2}{N} \sum_{k=1}^{N-1} (N-k)\rho(k) \right] \qquad (6.18)$$

The correlation $\rho(k)$ denotes the correlation between any pair of return vectors with lag k among them. If k equals 1, there are $2(N-1)$ such pairs. If k is $(N-1)$, however, there are only two pairs. Thus, a weighted sum of

---

* If it were cumulative returns, the variance would grow as N.

correlations of all possible lags would determine whether the impact of cross-sectional correlations contributes positively (negatively) to Equation 6.18, leading to a positive (negative) contribution to return effect and a negative (positive) effect on rebalancing alpha.

- In practice, the variance of cross-sectional returns is not constant for different periods, and cross-sectional serial correlation is also period specific. One would interpret variables on the right-hand side of Equation 6.18 as average over a sample horizon.

- The term inside the brackets would be the overall effect of serial correlations on the variance of arithmetic returns.

In summary, the results in this section provide a framework for analyzing the impact of cross-sectional serial correlations on the return effect of portfolio rebalancing. In later chapters, we return to this topic when we examine asset returns of similar volatilities.

6.3 APPROXIMATING RETURN EFFECTS OF LONG-SHORT PORTFOLIOS

Example 6.2 presents an example of a long-short portfolio with two assets. We found that the approximation of return effect by Equation 6.6 resulted in a significant error. In hindsight, we can point to two reasons why the formula in Equation 6.6 might be invalid for a long-short portfolio by reviewing the steps by which the formula is derived. First, in Equation 6.2, we use Taylor expansion of $(1 + g_i)^N$ at $g_i = \bar{g}$, the weighted average of geometric returns. When the portfolio weights are positive, $\bar{g}$ is within the range of g_i's. Hence, $g_i - \bar{g}$ should be small, and we expect reasonable accuracy in the Taylor expansion. However, this is not the case for long-short portfolios with negative weights. For a two-asset long-short portfolio, $\bar{g}$ is outside the interval formed by g_1 and g_2. This could cause the Taylor expansion to be less accurate.

The second and related reason is that for a long-short portfolio, $\text{var}(g)$ is no longer small, precisely because the weighted average $\bar{g}$ is outside the range. As we have seen, for a two-asset long-short portfolio, $\text{var}(g)$ is negative and could be significantly so. Since Equation 6.5 depends on $\text{var}(g)$ being small, this is another source of problems for long-short portfolios.

To mitigate these problems, we redefine a different average with alternative positive weights and then use the same approach in the case of long-only portfolios.

6.3.1 Two-Asset Long-Short Portfolios

We first consider the two-asset case, in which $w_1 < 0$, $w_2 > 0$, and $w_1 + w_2 = 1$. The return effect is negative, which can be expressed as

$$w_1 (1 + g_1)^N + w_2 (1 + g_2)^N = (1 + \bar{g} + e_r)^N \qquad (6.19)$$

Since $\bar{g}$ is now outside the interval formed by g_1 and g_2, we may not expand the left-hand side of Equation 6.19 around $\bar{g}$. On the other hand, note that g_2 is always between g_1 and $\bar{g}$. It turns out that g_2 is the weighted average of the other two measures, with positive weights. We have

$$g_2 = -\frac{w_1}{w_2} g_1 + \frac{1}{w_2} \bar{g} \triangleq \omega_1 g_1 + \bar{\omega} \bar{g} \qquad (6.20)$$

In Equation 6.20, we have defined the new weights ω_1 and $\bar{\omega}$. They are both positive and sum to 1. The variance based on these weights with respect to g_2 is then positive. We have

$$\mathrm{var}(g)\big|_2 = -\frac{w_1}{w_2}(g_1 - g_2)^2 + \frac{1}{w_2}(\bar{g} - g_2)^2 = -w_1(g_1 - g_2)^2 \qquad (6.21)$$

The symbol on the left-hand side denotes variance with respect to g_2. The new weighting scheme implies that we can now approximate the "return effect" of an alternative long-only portfolio with the weight ω_1 in Asset 1 and the weight $\bar{\omega}$ in another "asset"* with return $\bar{g}$. The result is

$$-\frac{w_1}{w_2}(1 + g_1)^N + \frac{1}{w_2}(1 + \bar{g})^N \approx (1 + g_2 + \tilde{e}_{r2})^N \qquad (6.22)$$

We have defined

$$\tilde{e}_{r2} = \frac{(N-1)}{2(1 + g_2)} \mathrm{var}(g)\big|_2 \qquad (6.23)$$

---

* This is not an asset formed by a linear combination of Assets 1 and 2. Nevertheless, the analysis follows through.

Note that Equation 6.23 is entirely analogous to Equation 6.6. As we have shown, they are valid approximations when the portfolio weights are positive.

However, Equation 6.23 is not yet the return effect of the original long-short portfolio defined in Equation 6.19. Now, by combining Equations 6.22 and 6.19, we derive an approximation to e_r in terms of $\tilde{e}_{r2}$. Multiplying Equation 6.22 by w_2 and expanding the right-hand side to the first two terms, we have

$$-w_1\left(1+g_1\right)^N +\left(1+\overline{g}\right)^N \approx w_2\left[\left(1+g_2\right)^N + N\tilde{e}_{r2}\left(1+g_2\right)^{N-1}\right] \qquad (6.24)$$

Similarly, Equation 6.19 is approximately

$$w_1\left(1+g_1\right)^N +w_2\left(1+g_2\right)^N \approx \left(1+\overline{g}\right)^N +Ne_r\left(1+\overline{g}\right)^{N-1} \qquad (6.25)$$

Adding Equations 6.24 and 6.25, and rearranging terms, we have

$$\tilde{e}_r \approx -\frac{\left(1+g_2\right)^{N-1}}{\left(1+\overline{g}\right)^{N-1}}w_2\tilde{e}_{r2} = \frac{\left(1+g_2\right)^{N-2}}{\left(1+\overline{g}\right)^{N-1}}\frac{\left(N-1\right)}{2}\left[-w_2\,\mathrm{var}\left(g\right)\big|_2\right] \qquad (6.26)$$

In the second expression, we have substituted Equation 6.23 for $\tilde{e}_{r2}$. Finally, using Equation 6.21, we have

$$\tilde{e}_r = \frac{\left(1+g_2\right)^{N-2}}{\left(1+\overline{g}\right)^{N-1}}\frac{\left(N-1\right)}{2}\left[w_1w_2\left(g_1-g_2\right)^2\right] = \frac{\left(1+g_2\right)^{N-2}}{\left(1+\overline{g}\right)^{N-2}}\left[\frac{\left(N-1\right)}{2\left(1+\overline{g}\right)}\,\mathrm{var}\left(g\right)\right] \qquad (6.27)$$

In the second expression, we have used the original definition of $\mathrm{var}\left(g\right)$— the weighted variance with the original weights.

We make several remarks about this result.

- This approximation is consistent with the fact that the return effect of a two-asset long-short portfolio is always negative.

- We recognize that the term in the brackets is of the same form as the return effect of a long-only portfolio. Thus, the return effect of the BH long-short portfolio is obtained by scaling the formula for long-only return effect by a ratio of two powers. The numerator is a power of the return of the long asset (Asset 2 in our case), and the denominator is a power of the average return.

- When $g_1 < g_2$, we have $g_2 < \bar{g}$. In other words, when the return of the short asset is less than the return of the long asset, the return of the portfolio exceeds that of the long asset. Consequently, the ratio of two powers of two respective returns will be less than 1.

Example 6.3: We revisit the two-asset long-short portfolio considered previously in Example 6.2. The portfolio is long 150% in stocks and short 50% in bonds. Hence, $g_2 = 10.48\%$, and then, the ratio in Equation 6.27 is

$$\frac{\left(1+g_2\right)^{N-2}}{\left(1+\bar{g}\right)^{N-2}} = 0.3594$$

The number N is 45. Multiplying this ratio to the original approximation of -4.13% in Table 6.3, we obtain a better approximation of -1.48%, both of which are listed in Table 6.4.

6.3.2 General Long-Short Portfolios

For the general case with more than two assets, a similar approximation exists. We shall first state the result and then prove a special case and leave the general case as an exercise.

We group assets into two groups: one with long assets and the other with short assets. Define the average geometric mean of long assets by

$$\bar{g}_p = \frac{1}{w_P} \sum_{w_i>0} w_i g_i, \quad \text{with } w_P = \sum_{w_i>0} w_i \tag{6.28}$$

Then, the return effect is simply

$$\tilde{e}_r = \frac{\left(1+\bar{g}_p\right)^{N-2}}{\left(1+\bar{g}\right)^{N-2}}\left[\frac{(N-1)}{2\left(1+\bar{g}\right)} \operatorname{var}(g)\right] \tag{6.29}$$

TABLE 6.4 Two Approximations for the Return
Effect of a Long-Short Portfolio

	Return Effect
Actual	−1.75%
Approximation	−4.13%
Alternative	−1.48%

The scaling ratio is a power of the average mean of long assets over a power of the overall mean. The generalized variance in the bracket term is with respect to the original long and short weights.

The special case we shall prove is one in which there is only one short position and the rest are long positions. In other words, we have $w_1 < 0$, $w_i > 0, i = 2, \ldots, M$. Then, the sum of positive weights is

$$w_P = \sum_{i=2}^{M} w_i = 1 - w_1 \tag{6.30}$$

And the average return for long assets is

$$\bar{g}_p = \frac{1}{w_P} \sum_{i=2}^{M} w_i g_i = \frac{1}{w_P} \left(\bar{g} - w_1 g_1 \right) \tag{6.31}$$

Grouping the long assets together, we have an approximation for their return effect:

$$\frac{1}{w_P} \sum_{i=2}^{M} w_i \left(1 + g_i\right)^N = \left(1 + \bar{g}_p + e_{r,p}\right)^N \tag{6.32}$$

We have

$$e_{r,p} \approx \frac{(N-1)}{2(1+\bar{g}_p)} \operatorname{var}\left(g_p\right) \tag{6.33}$$

In the approximation, $\operatorname{var}\left(g_p\right)$ is the variance of returns restricted to assets with long positions. We have

$$\operatorname{var}\left(g_p\right) = \frac{1}{w_P} \sum_{i=2}^{M} w_i \left(g_i - \bar{g}_p\right)^2 = \frac{1}{w_P} \sum_{i=2}^{M} w_i \left(g_i - \bar{g}\right)^2 - \frac{w_1^2}{w_P} \left(g_1 - \bar{g}\right)^2 \tag{6.34}$$

The second step in the equation converts the variance restricted to long positions to terms in the variance for all assets.

Grouping the short asset and the average return "asset" together, we have

$$\frac{1}{w_P} \left(1 + \bar{g}\right)^N - \frac{w_1}{w_P} \left(1 + g_1\right)^N \tag{6.35}$$

According to Equation 6.31, the weighted average of two returns involved also equals $\bar{g}_p$. Therefore, we can approximate Equation 6.35 by

$$\frac{1}{w_P}\left(1+\bar{g}\right)^N - \frac{w_1}{w_P}\left(1+g_1\right)^N = \left(1+\bar{g}_p + e_{r,n}\right)^N \tag{6.36}$$

$$e_{r,n} \approx \frac{(N-1)}{2\left(1+\bar{g}_p\right)} \text{var}\left(g_n\right) \tag{6.37}$$

The variance $\text{var}\left(g_n\right)$ is given by

$$\text{var}\left(g_n\right) = \frac{1}{w_P}\left(\bar{g}-\bar{g}_p\right)^2 - \frac{w_1}{w_P}\left(g_1-\bar{g}_p\right)^2 = -\frac{w_1}{w_P^2}\left(g_1-\bar{g}\right)^2 \tag{6.38}$$

Combining Equations 6.32 and 6.36 yields

$$\sum_{i=1}^{M} w_i\left(1+g_i\right)^N \approx \left(1+\bar{g}\right)^N + w_p\left(1+\bar{g}_p + e_{r,p}\right)^N - w_p\left(1+\bar{g}_p + e_{r,n}\right)^N \tag{6.39}$$

$$\approx \left(1+\bar{g}\right)^N + w_p N\left(1+\bar{g}_p\right)^{N-1}\left(e_{r,p} - e_{r,n}\right)$$

Equation 6.39 can be further approximated by

$$\sum_{i=1}^{M} w_i\left(1+g_i\right)^N \approx \left[1+\bar{g} + \frac{\left(1+\bar{g}_p\right)^{N-1}}{\left(1+\bar{g}\right)^{N-1}} w_p\left(e_{r,p} - e_{r,n}\right)\right]^N \tag{6.40}$$

This implies that the return effect of the overall portfolio can be approximated by

$$e_r \approx \frac{\left(1+\bar{g}_p\right)^{N-1}}{\left(1+\bar{g}\right)^{N-1}} w_p\left(e_{r,p} - e_{r,n}\right) \tag{6.41}$$

Using Equations 6.33 and 6.37, we have

$$w_p\left(e_{r,p} - e_{r,n}\right) \approx \frac{(N-1)w_p}{2\left(1+\bar{g}_p\right)}\left[\text{var}\left(g_p\right) - \text{var}\left(g_n\right)\right] \tag{6.42}$$

From Equations 6.34 and 6.38, we find

$$w_p\left[\text{var}(g_p)-\text{var}(g_n)\right]=\sum_{i=1}^{M}w_i\left(g_i-\bar{g}\right)^2=\text{var}(g) \tag{6.43}$$

We recover the variance of returns of all assets in the portfolio. The final equation is the desired result:

$$e_r\approx\frac{\left(1+\bar{g}_p\right)^{N-2}}{\left(1+\bar{g}\right)^{N-2}}\left[\frac{(N-1)}{2\left(1+\bar{g}\right)}\text{var}(g)\right] \tag{6.44}$$

For a long-only portfolio, the two averages are the same. Therefore, Equation 6.44 simplifies to Equation 6.6. In the two-asset case, $\bar{g}_p=g_2$ and $\text{var}(g)$ is negative. For the general case, $\text{var}(g)$ could be either positive or negative, depending on portfolio weights and geometric returns of individual assets. Hence, for long-short portfolios, we can use Equation 6.44 to determine approximately whether the return effect is positive or negative.

For one particular class of long-short portfolios with only one asset with long position, we know from Problem 4.7 that the return effect is always negative. The two-asset long-short portfolio is a special case. It turns out that $\text{var}(g)$ is indeed negative. Thus, the approximation is consistent in terms of the sign. We leave the proof as an exercise.

PROBLEMS

6.1. For a two-asset long-short portfolio, in which $w_1<0$, $w_2>0$, and $w_1+w_2=1$, prove that g_2 is always between g_1 and $\bar{g}$.

6.2. Prove the second step in Equation 6.34.

6.3. Prove the result in Equation 6.44 for a general long-short portfolio.

6.4. For long-short portfolios with M assets, we have $w_i<0, i=1,\cdots,M-1$, and $w_M>0$. Prove that $\text{var}(g)<0$.

CHAPTER 7

Analysis of Rebalancing Alpha

H AVING ANALYZED BOTH THE volatility and return effects of port-folio rebalancing in Chapters 5 and 6, we now combine the results together to derive an approximation of rebalancing alpha. As in the previous analysis, we first gain insights by examining the case of two-asset portfolios.

7.1 REBALANCING ALPHA OF TWO-ASSET PORTFOLIOS

Using approximations of volatility (Equation 5.11) and return effects (Equations 6.6 and 6.7), we have

$$\tilde{\alpha} \approx \tilde{e}_v - \tilde{e}_r = \frac{1}{2} w_1 w_2 \sigma_{1\_2}^2 - \frac{(N-1)}{2} w_1 w_2 \left(g_1 - g_2 \right)^2 \tag{7.1}$$

In Equation 7.1, we have omitted the term $\left(1 + \bar{g}\right)$ from the return effect. In addition, we note that this expression is only appropriate for long-only portfolios.

7.1.1 Pairwise t-Statistics

Factoring the portfolio weights and the variance term leads to

$$\tilde{\alpha} = \frac{1}{2} w_1 w_2 \sigma_{1\_2}^2 \left(1 - \tilde{t}_{1\_2}^2 \right) \tag{7.2}$$

We have identified the term inside the brackets as a variant of t-statistics of relative return difference between the two assets:

$$\tilde{t}_{1\_2} \triangleq \frac{\sqrt{N-1}\left(g_1 - g_2\right)}{\sigma_{1\_2}} \tag{7.3}$$

Equation 7.3 is different from the conventional t-statistics in three aspects. First, we have geometric means instead of arithmetic means. Second, our estimate of volatility is not based on the unbiased estimator of sample variance. Third, the factor involving the sample size is $\sqrt{N-1}$ instead of $\sqrt{N}$. For the time being, we shall use Equation 7.3 as a measure to analyze empirical data. We shall make appropriate adjustments to it shortly, when we analyze the statistical properties of rebalancing alpha.

Equation 7.2 is a succinct and important result. It is worth exploring its full implications. For long-only portfolios, since the weights are positive, the sign of rebalancing alpha depends on the magnitude of the t-statistic. We have

$$\tilde{\alpha} \begin{cases} > 0, & \left|\tilde{t}_{1\_2}\right| < 1 \\ = 0, & \left|\tilde{t}_{1\_2}\right| = 1 \\ < 0, & \left|\tilde{t}_{1\_2}\right| > 1 \end{cases} \tag{7.4}$$

When the absolute value of the t-statistic is less than 1, the rebalancing alpha is positive, and when the absolute value of the t-statistic is greater than 1, the rebalancing alpha is negative. When the t-statistic is 0, which occurs when the two assets have the same sample geometric returns, the value added of portfolio rebalancing is the highest. In other words, if there is no eventual winner or loser, then selling temporary winners and buying temporary losers yields the greatest possible alpha for a long-only two-asset portfolio.

Suppose one of the two assets is the risk-free asset; then, the t-statistic becomes the term $\sqrt{N-1}$ times the Sharpe ratio of the risky asset, because the difference of two returns is the excess return of the risky asset, and the volatility of return difference is essentially the volatility of the risky asset. It is in fact the t-statistic for the positive excess return of the risky asset. While a t-statistic of 2 is normally the threshold of statistical significance for the excess return, a t-statistic of 1 is sufficient for the rebalancing alpha to be negative and for the buy-and-hold (BH) portfolio to outperform the fixed-weight (FW) portfolio.

Other factors in Equation 7.2 have a direct impact on the size of the rebalancing alpha, not on the sign. The rebalancing alpha is directly proportional to the variance of return differences.* Of course, this linear relationship works both ways: when the rebalancing alpha is either positive or negative. The product of the two weights varies between the minimum value of 0 (when the weight of one asset is 0) and the maximum value of 0.25 (when the weights of both assets are equal and at 0.5). Therefore, an equally weighted portfolio is one that achieves either maximum positive rebalancing alpha or minimum negative rebalancing alpha. Numerically, if the t-statistic is 0, the weights are 50/50, and the volatility of return differential is 20%, then the rebalancing return is 50 basis points (bp), or 0.5% $\left[= 0.5 \cdot 0.5 \cdot 0.5 \cdot (0.2)^2 \right]$.

We also note that the rebalancing alpha does not depend explicitly on mean reversion of asset returns. Of course, mean reversion would lessen return effect, so it implicitly contributes to rebalancing alpha. It remains to be seen whether rebalancing alpha can be positive if the net effect of serial correlations is zero. We shall return to this question later in the book.

Example 7.1: We consider rebalancing alpha of four 50/50 asset allocation portfolios that are studied in Chapter 5 and 6 in terms of volatility and return effects. They are 50/50 stock/cash, 50/50 stock/bond, 50/50 commodity/stock, and 50/50 commodity/bond portfolios based on annual returns from 1970 to 2014.

Table 7.1 lists the t-statistics for the four pairs of assets in the four portfolios, together with the difference of returns and the volatility of return differences. The t-statistic of the cash/stock pair is nearly 2, while the t-statistic of the stock/bond pair is about 1. The other two t-statistics are below 1. The approximations of rebalancing alpha given in Table 7.1 are consistent with the magnitudes of t-statistics, and they are close to the actual rebalancing alpha except for the 50/50 cash/stock portfolio.

---

* This is apparently the reason behind the unflattering term *volatility pumping* being associated with portfolio rebalancing.

TABLE 7.1   *t*-Statistics for Four Asset Allocation Portfolios, with Actual and Approximate Rebalancing Alpha

	50/50 Cash/Stock	50/50 Stock/Bond	50/50 Commodity/Stock	50/50 Commodity/Bond
$|g_1 - g_2|$	5.32%	2.90%	2.45%	0.45%
σ_{1_2}	17.63%	18.93%	31.51%	27.33%
$|\tilde{t}_{1\_2}|$	2.00	1.02	0.52	0.11
α	−1.17%	−0.01%	0.91%	0.92%
$\tilde{\alpha}$	−0.83%	0.06%	0.85%	0.92%

7.1.2 Probability of Positive Rebalancing Alpha

If we approximate the geometric means by arithmetic means, we have

$$\tilde{\alpha} \approx \frac{1}{2} w_1 w_2 \left[\sigma_{1\_2}^2 - (N-1)\mu_{1\_2}^2 \right] \tag{7.5}$$

The term μ_{1_2} denotes the difference between the two arithmetic means. This approximation is reasonable when the variances of both assets are the same or small relative to the arithmetic means. Equation 7.5 expresses rebalancing alpha in terms of sample mean and sample variance of the return difference between the two assets. However, recall that the definition of $\sigma_{1\_2}^2$ has N in the denominator instead of $(N-1)$ as in the definition of variance s^2. We rewrite Equation 7.5 as

$$\tilde{\alpha} \approx \frac{1}{2} w_1 w_2 \left[\frac{(N-1)}{N} s_{1\_2}^2 - (N-1)\bar{\mu}_{1\_2}^2 \right] \tag{7.6}$$

With Equation 7.6, it is possible to determine certain statistical properties of approximated rebalancing alpha given the underlying return distribution of the two assets and the investment horizon. So far, we have avoided prescribing statistical distributions of asset returns. Now, we assume that the return difference of the two assets is normal and that over the return horizon, it is independent and identically distributed. This allows us to derive the probability of positive rebalancing alpha.

It is well known that the sample mean follows a normal distribution:

$$\bar{\mu}_{1\_2} \sim N\left(\mu_{1\_2}, \frac{\sigma_{1\_2}^2}{N} \right) \tag{7.7}$$

and the sample variance follows a scaled chi-square distribution with $(N-1)$ degrees of freedom:

$$(N-1)\frac{s_{1\_2}^2}{\sigma_{1\_2}^2} \sim \chi_{N-1}^2 \tag{7.8}$$

Note that μ_{1_2} and σ_{1_2} denote population mean and population standard deviation, respectively.

We consider the probability of positive rebalancing alpha when the two assets have the same expected return, i.e. $\mu_{1\_2} = 0$. By Equation 7.6, we have

$$\tilde{\alpha} \approx \frac{1}{2} w_1 w_2 s_{1\_2}^2 \frac{N-1}{N}\left(1-\frac{N\overline{\mu}_{1\_2}^2}{s_{1\_2}^2}\right) = \frac{1}{2} w_1 w_2 s_{1\_2}^2 \frac{N-1}{N}\left(1-t_{1\_2}^2\right) \tag{7.9}$$

Now, t_{1_2} is the actual t-statistic, and under the aforementioned assumptions, it follows a t-distribution with $N-1$ degrees of freedom. The probability of $\tilde{\alpha} > 0$ is the cumulative probability of $|t_{1\_2}| < 1$. For example, if $N = 5$, i.e. the number of periods is 5, the probability of $\tilde{\alpha} > 0$ is about 63%. If $N = 10$, the probability increases slightly to 66%, or a chance of 2 out 3. Figure 7.1 plots the probability of positive rebalancing alpha as a function of investment horizon. When the investment period equals 2, the probability is 0.5. As the investment period lengthens, the probability increases, at first rather quickly, and then slowly converges to a value of 0.6827. This is because the t-distribution in question approximates to the standard normal distribution as the degrees of freedom become large.

When the expected returns of the two assets are not equal, the ratio of sample statistics follows a non-central t-distribution with $N-1$ degrees of freedom. We have

$$\tilde{\alpha} \approx \frac{1}{2} w_1 w_2 s_{1\_2}^2 \frac{N-1}{N}\left[1-T_{N-1}^2(\delta)\right] \tag{7.10}$$

The non-centrality parameter δ is given by

$$\delta = \frac{\sqrt{N}\mu_{1\_2}}{\sigma_{1\_2}} \tag{7.11}$$

This parameter resembles a "population t-statistic," and it shares a similarity with an "annualized" return/risk ratio of asset returns. For example,

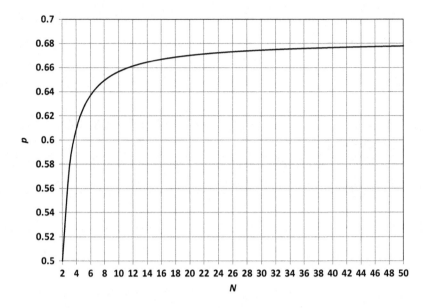

FIGURE 7.1 Probability of positive rebalancing alpha as a function of N, with equal expected returns.

if a portfolio is rebalanced monthly for 12 months, and μ_{1_2} and σ_{1_2} are monthly expected return and standard deviation, respectively, then δ is the annualized Sharpe ratio. If a portfolio is rebalanced annually for 10 years, and μ_{1_2} and σ_{1_2} are yearly expected return and standard deviation, respectively, then δ is the Sharpe ratio for a decade.

The probability of positive rebalancing alpha is then given by the probability of $\left| T_{N-1}(\delta) \right| < 1$.

Example 7.2: We consider the hypothetical probability of positive rebalancing alpha of four asset allocation portfolios in Example 7.1. We shall assume that the underlying return distributions are normal, independent, and identically distributed, with expected return and standard deviation that coincide with the sample parameters. Since the probability is independent of portfolio weights, they apply to any long-only portfolios consisting of these pairs of assets.

Table 7.2 lists the non-centrality parameters of the four cases and the associated probability of positive rebalancing alpha. For commodity/bond portfolios, since the difference of expected returns of

TABLE 7.2   Non-Centrality Parameter and Probability of Positive Rebalancing
Alpha for Four Two-Asset Portfolios

	Cash/Stock	Stock/Bond	Commodity/Stock	Commodity/Bond
μ_{1_2}	5.32%	2.90%	2.45%	0.45%
σ_{1_2}	17.63%	18.93%	31.51%	27.33%
δ	2.03	1.03	0.52	0.11
$p(\tilde{\alpha} > 0)$	0.15	0.46	0.62	0.67

the two assets is small, the non-centrality parameter is close to zero.
As a result, the probability is similar to that given in Example 7.1
with $N = 45$. For stock/bond portfolios, the non-centrality parameter is close to 1, and the probability is about 0.46. Hence, over any
given 45-year period, the chance of getting positive rebalancing
alpha is slightly less than 50%. For cash/stock portfolios, the non-
centrality parameter is close to 2, and the probability drops to 0.15.
While small, this probability is far from zero. Hence, there is still a
decent chance that a FW cash/stock portfolio could outperform its
BH cousin after 45 years.

Example 7.3: We consider portfolios with stocks and bonds under
the following return assumptions. The expected returns for stocks
and bonds are 2% and 6%, and their annual volatilities are 5% and
15%, respectively. These lead to an equal Sharpe ratio of 0.4 and
equity premium over bonds of $\mu_{1\_2} = 4\%$. We investigate three
cases of return correlation between the two assets: $\rho = -0.3, 0, 0.3$,
which implies three levels of volatility of the return differentials:
$\sigma_{1\_2} = 17.18\%, 15.81\%, 14.32\%$. Of course, we assume that returns are
independent and identically distributed.

   With varying investment horizon N, we calculate the probability
of $|T_{N-1}(\delta)| < 1$, and the results are displayed in Figure 7.2 for three
cases of return correlations. These are quite different from the result
in Figure 7.1, where the expected returns are the same for the two
assets.

- The higher the correlation, the lower the probability of positive
  rebalancing alpha. This is because a higher correlation leads to
  lower volatility of return differences between the two assets,
  which reduces volatility effects.

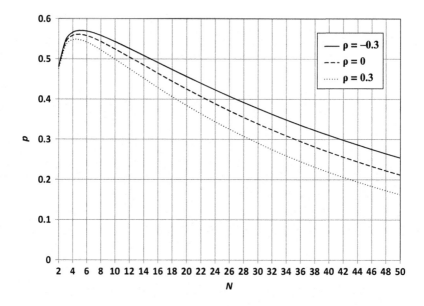

FIGURE 7.2   Probability of positive rebalancing alpha for stock/bond portfolios as a function of investment horizon for three different correlations between the two assets.

- For each correlation, the probability of positive rebalancing alpha increases first. The maximum probability seems to be around 5 years. However, it eventually decreases as N gets large. When N reaches 50 years, the probability is quite low. For instance, if the correlation is 0, the probability of positive alpha when the investment horizon is 50 years is around 0.2.

7.1.3 Expected Value and Standard Deviation of Rebalancing Alpha

The derivation of the probability of positive rebalancing alpha requires specification of the underlying return distributions. There is no such requirement if we are interested in the expected value of rebalancing alpha. However, we take advantage of the known results based on the assumption of normal distribution to simplify the derivation.

From Equation 7.6, we have

$$E(\tilde{\alpha}) \approx \frac{1}{2} w_1 w_2 (N-1) \left[\frac{E(s_{1\_2}^2)}{N} - E(\bar{\mu}_{1\_2}^2) \right] \tag{7.12}$$

Since $\bar{\mu}_{1\_2}$ is a normal distribution given by Equation 7.7, we have

$$E\left(\bar{\mu}_{1\_2}^2\right) = \mu_{1\_2}^2 + \frac{\sigma_{1\_2}^2}{N} \tag{7.13}$$

In addition, since $s_{1\_2}^2$ follows the scaled chi-square distribution given by Equation 7.8, we have

$$E\left(s_{1\_2}^2\right) = \sigma_{1\_2}^2 \tag{7.14}$$

Substituting Equations 7.13 and 7.14 into Equation 7.12 yields

$$E(\tilde{\alpha}) \approx -\frac{1}{2} w_1 w_2 \left(N-1\right) \mu_{1\_2}^2 \tag{7.15}$$

We make several remarks about this result.

- The result is valid without the assumption of underlying normal distribution. However, the returns are independent and identically distributed in different periods.

- For a long-only portfolio of two assets, the expected rebalancing alpha is always negative when the expected returns of the two assets are different. Consequently, rebalancing alpha is likely to be negative when the returns are serially independent. Therefore, it seems that for an FW portfolio to outperform a BH portfolio, some degree of mean reversion or negative serial correlations are needed.

- When the expected returns are the same, the expected rebalancing alpha is zero. Recall that the probability of positive rebalancing alpha is greater than 0.5. The implication of these two results is the distribution of rebalancing alpha must not be symmetric.

The variance of rebalancing alpha is given by

$$\mathrm{var}(\tilde{\alpha}) \approx \left[\frac{1}{2} w_1 w_2 \left(N-1\right)\right]^2 \left[\frac{\mathrm{var}\left(s_{1\_2}^2\right)}{N^2} + \mathrm{var}\left(\bar{\mu}_{1\_2}^2\right)\right] \tag{7.16}$$

When $s_{1\_2}^2$ follows the scaled chi-square distribution given by Equation 7.8, we have

$$\operatorname{var}\left(s_{1\_2}^2\right) = \frac{2\sigma_{1\_2}^4}{N-1} \tag{7.17}$$

When $\bar{\mu}_{1\_2}$ is a normal distribution given by Equation 7.7, we have

$$\operatorname{var}\left(\bar{\mu}_{1\_2}^2\right) = \frac{2\sigma_{1\_2}^4}{N^2} + \frac{4\mu_{1\_2}^2\sigma_{1\_2}^2}{N} \tag{7.18}$$

Substituting Equations 7.17 and 7.18 into Equation 7.16 and taking the square root gives

$$\operatorname{std}\left(\tilde{\alpha}\right) \approx w_1 w_2 \sigma_{1\_2}^2 \sqrt{\frac{(N-1)}{2N}} \sqrt{1 + \frac{2(N-1)\mu_{1\_2}^2}{\sigma_{1\_2}^2}} \tag{7.19}$$

Several features are worth noting.

- The standard deviation is proportional to the product of weights and variance of return differences between the two assets.

- When the expected returns of the two assets are the same, i.e. $\bar{\mu}_{1\_2} = 0$, the standard deviation of rebalancing alpha is bounded as a function of investment period N. When N gets large, it increases slightly and converges to $w_1 w_2 \sigma_{1\_2}^2 / \sqrt{2}$.

- However, when the expected returns are not the same, the standard deviation of rebalancing alpha increases more rapidly as the investment period lengthens. When N gets large, it is approximately given by $\sqrt{N-1} w_1 w_2 |\mu_{1\_2}| \sigma_{1\_2}$. Comparing this result with Equation 7.15, we note that the expected rebalancing alpha deteriorates as a linear function of N, while the standard deviation increases as a square root function of N.

7.1.4 Distribution of Rebalancing Alpha

Theoretically, it is possible to derive the distribution of rebalancing alpha from Equation 7.6 under the assumption of an independent and identically distributed normal distribution with zero means. This is because the rebalancing alpha is then proportional to the difference of two chi-square distributions, which are also independent. Unfortunately, the resulting random variable does not seem to have an explicit analytic density function. Therefore, we resort to numerical simulations to analyze the distribution of rebalancing alpha. In addition, we can use simulation to gauge the error of using Equation 7.6 in approximating the actual rebalancing alpha.

We draw returns of two assets from a bivariate normal distribution of given means and a covariance matrix. The return volatilities resemble those of bonds and stocks, and they are 5% and 15%, respectively. We also assume that the correlation is zero between the two assets. In terms of expected returns, we study two cases. In the first case, the expected returns are the same at 2%. In the second case, the expected return of stocks is 6%, and that of bonds remains at 2%. It should be noted that the second assumption is more plausible. We study the first case because it is a special case where the expected rebalancing alpha is zero. We simulate independent annual returns up to 20 years. Based on a sample size of 500,000 scenarios, we analyze statistical properties and distributions of actual rebalancing alpha and approximations. Here are the assumptions.

$$\Sigma = \begin{pmatrix} 0.0025 & 0 \\ 0 & 0.0225 \end{pmatrix}, \mu = \begin{cases} \left(2\% \quad 2\%\right)', \text{ Case 1} \\ \left(2\% \quad 6\%\right)', \text{ Case 2} \end{cases} \quad (7.20)$$

Figure 7.3 plots the mean and standard deviation of rebalancing alpha under Case 1. As indicated by previous analysis, the average alpha is close to zero. This is true for both actual alpha and the approximation given in Equation 7.5. The standard deviation increases as the investment horizon lengthens. When $N = 2$, it is about 30 bp, and it increases to about 40 bp when $N = 5$. As N increases further, the standard deviation seems to converge. The approximated value in this case slightly overstates the actual value.

In Figure 7.4, we plot the skewness and kurtosis of the rebalancing alpha based on the same 500,000 simulations. We note that the most significant feature of the graphs is the negative skewness and excess kurtosis of rebalancing alpha. In other words, the rebalancing alpha under the return assumption has fat tail risks. The skewness is close to zero when $N = 2$, in which there is only one rebalancing. As the investment horizon lengthens, the skewness becomes more negative, approaching a limiting value of −2. The skewness based on the approximation of rebalancing alpha is slightly

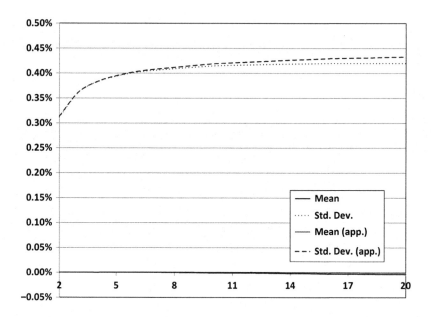

FIGURE 7.3   Actual and approximated mean and standard deviation of rebalancing alpha for Case 1.

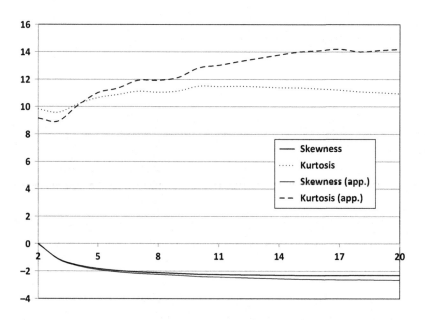

FIGURE 7.4   Actual and approximated skewness and kurtosis of rebalancing alpha for Case 1.

off. The error in kurtosis based on the approximation is relatively larger. For the actual alpha, the kurtosis is near 10 when $N = 2$, and it increases to about 11 when N gets large. However, for the approximation, the kurtosis starts lower than 10 but grows much higher, to 14.

We present similar simulation results for Case 2, in which the expected returns of two assets are not the same. This results in negative expected rebalancing alpha.

The mean and standard deviation of actual and approximated alpha are shown in Figure 7.5. Both grow in magnitude as the investment horizon lengthens. However, we observe that unlike in Case 1, the differences of these two statistics in Case 2 are rather large. This is due to using arithmetic means for geometric means in the approximation Equation 7.5. An alternative approximation, based on geometric means and the additional factor $1 + \bar{g}$ in the denominator for the return effect, would improve the results substantially. In Figure 7.5, we also plot the mean and standard deviation based on this alternative approximation (labelled app2). They are much closer to the actual values.

- While the alternative approximation has much better accuracy, it is not analytically tractable due to the presence of geometric means.

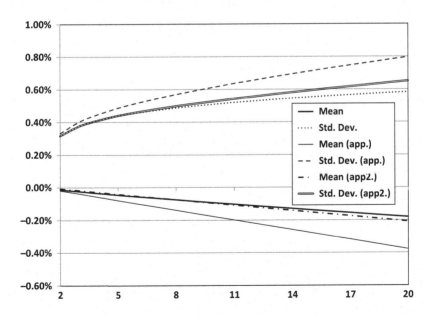

FIGURE 7.5 Actual and approximated mean and standard deviation of rebalancing alpha for Case 2.

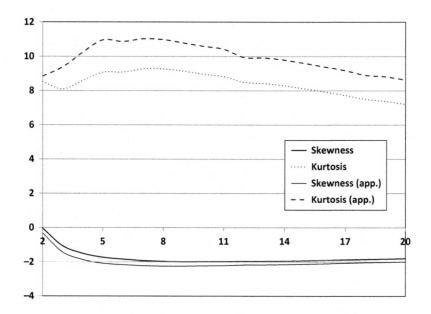

FIGURE 7.6 Actual and approximated skewness and kurtosis of rebalancing alpha for Case 2.

Figure 7.6 provides the skewness and kurtosis for the actual and approximated rebalancing alpha of Case 2. As in Case 1, the approximated skewness is slightly more negative than the actual value, while the approximated kurtosis is higher than the actual value.

7.2 REBALANCING ALPHA OF GENERAL PORTFOLIOS

We now consider rebalancing alpha of a portfolio with M assets by using the results of volatility and return effects from Chapters 5 and 6, respectively.

7.2.1 Pairwise Decomposition of Rebalancing Alpha

Combining Equations 5.15 and 6.9 and factoring the common terms, we have

$$\tilde{\alpha} = \frac{1}{2}\sum_{i<j}^{M} w_i w_j \left[\sigma^2_{i\_j} - (N-1)(g_i - g_j)^2 \right] \qquad (7.21)$$

Similarly, we have omitted the term $(1+\bar{g})$ from the return effect in Equation 7.21. We note that this is a direct generalization of Equation 7.1. We define a pairwise t-statistic by

$$\tilde{t}_{i\_j} \triangleq \frac{\sqrt{N-1}\left(g_i - g_j\right)}{\sigma_{i\_j}} \tag{7.22}$$

Then, we have a generalization of Equation 7.2:

$$\tilde{\alpha} = \frac{1}{2} \sum_{i<j}^{M} w_i w_j \sigma_{i\_j}^2 \left(1 - \tilde{t}_{i\_j}^2\right) \tag{7.23}$$

Intuitively, rebalancing alpha of a portfolio is simply the sum of rebalancing alphas of all distinct pairs. Notice that the portfolio weights are separable from the term that consists of volatilities, investment returns, and investment horizon, which are "intrinsic" to asset returns and investment period.

We make several remarks about this general result. First, rebalancing alpha for a long-only portfolio will be positive if rebalancing alphas of all pairs are non-negative and there is at least one positive pair. A special case in this regard, albeit a theoretical one, is when all geometric returns are the same. Then, the return effect vanishes, and rebalancing alpha is just the volatility effect. Second, suppose there are some pairs with negative rebalancing alphas. If one can judiciously choose portfolio weights that exclude those pairs, it will lead to a positive rebalancing alpha. Of course, this requires forecasting of pairwise rebalancing alpha, which is hard to do in reality. Third, we recall that in the two-asset case, an equally weighted portfolio gives extreme rebalancing alpha. There is a similar result in the general case. It can be shown that if the term $\left[\sigma_{i\_j}^2 - (N-1)(g_i - g_j)^2\right]$ is the same for all pairs, then an equally weighted portfolio achieves extreme rebalancing alpha, either positive or negative.

The matrix form of rebalancing alpha is given by combining Equations 5.16 and 6.11. We have

$$\tilde{\alpha} = \frac{1}{4} \mathbf{w}' \left[\Sigma_v - (N-1)\Sigma_g\right] \mathbf{w} \tag{7.24}$$

The matrices Σ_v and Σ_g were defined previously in Chapters 5 and 6. We note that Equation 7.24 is similar to the variance of a portfolio in terms of the weight vector and a covariance matrix. The difference is that in the case of rebalancing alpha, the matrix is not a covariance matrix. Rather, it is a linear combination of two "distance matrices" between distinct pairs.

- A distance matrix consists of elements that are related to pairwise distances between points in a certain vector space. For Σ_g, the space is a one-dimensional line, and it can be proved that Σ_g is singular with a rank of at most 3.

- For Σ_v, the vector space consists of random variables and the distance is defined as variance between two random variables. In general, it is not singular. But it is not positive definite either.

7.2.2 An Alternative Decomposition of Rebalancing Alpha

Portfolio rebalancing can also be thought of as a combination of pairwise rebalancing between individual assets and the FW portfolio. From Equations 5.18 and 6.6, we have

$$\tilde{\alpha} = \frac{1}{2}\sum_{i=1}^{M} w_i \sigma^2_{i-\text{FW}} - \frac{(N-1)}{2(1+\bar{g})}\text{var}(g) \tag{7.25}$$

Omitting the term $(1+\bar{g})$ and expanding the term $\text{var}(g)$, we have

$$\tilde{\alpha} = \frac{1}{2}\sum_{i=1}^{M} w_i \left[\sigma^2_{i-\text{FW}} - (N-1)(g_i - \bar{g})^2\right] \tag{7.26}$$

The term in the brackets is approximately the pairwise rebalancing alpha between asset i and the FW portfolio. The slight difference lies in the fact that $\bar{g}$ is not the geometric return of the FW portfolio g_{FW}.

7.2.3 Expectation of Portfolio Rebalancing Alpha

Suppose the population distribution of the asset returns is a multivariate normal distribution:

$$\mathbf{r} \sim N(\boldsymbol{\mu}, \boldsymbol{\Sigma}) \tag{7.27}$$

Then by Equation 7.15, the expected pairwise rebalancing alpha in Equation 7.21 is given by $-w_i w_j (N-1) \mu_{i-j}^2 / 2$. Summing over all pairs, we have

$$E(\tilde{\alpha}) = -\frac{(N-1)}{2} \sum_{i<j}^{M} w_i w_j \mu_{i-j}^2 = -\frac{(N-1)}{2} \sum_{i<j}^{M} w_i w_j (\mu_i - \mu_j)^2 \qquad (7.28)$$

The last expression can be rewritten as the variance of expected returns of individual assets. We have

$$E(\tilde{\alpha}) = -\frac{(N-1)}{2} \operatorname{var}(\mu) \qquad (7.29)$$

This result is an extension of Equation 7.15, which again shows that the expected rebalancing alpha of a long-only portfolio is negative unless all assets have the same expected return.

- The variance of rebalancing alpha for a general long-only portfolio is much more involved. Its derivation probably depends on the distribution of the sample covariance matrix.

7.2.4 Rebalancing Alpha—S&P 500 Sector Portfolios

So far, we have used asset allocation portfolios with stocks, bonds, and commodities for empirical examinations of portfolio rebalancing. The benefit of using portfolios with only two or three asset classes is that it is simple to see the trade-off between volatility and return effects of a few pairs of assets. The drawback is that expected returns of different asset classes are potentially very different, due to either different levels of risk or different reactions to macroeconomic regimes. When the return difference is large, the probability of negative rebalancing alpha is high. Therefore, portfolio rebalancing of asset allocation portfolios, while indispensable for managing risks, might not lead to positive rebalancing alpha.

In contrast, portfolio rebalancing among similar assets that are from the same segments of capital markets has a better chance of achieving positive rebalancing alpha. This assertion, while not yet verified, has two potential arguments. First, the expected returns of these assets should show smaller differences because of similar economic exposures and similar levels of risk. Second, as they belong to the same market, it is possible that returns of these assets would exhibit some degree of cross-sectional mean reversion. Both arguments bolster the probability of low return effect of portfolio rebalancing (for long-only portfolios) and positive rebalancing alpha.

We thus introduce our second empirical example of equity portfolios with sector allocations. We use 10 sectors in the S&P 500 index to build portfolios and compare the performance of FW portfolios and BH portfolios. In the next few chapters, we will extend the examination to other portfolios in stocks, bonds, and commodities.

As we have shown earlier, portfolio rebalancing alpha depends on the pairwise rebalancing alphas of the 10 sectors and initial portfolio weights. Therefore, the first step is to analyze the returns and risks of all distinct sector pairs.

Table 7.3 displays geometric return, volatility, and Sharpe ratio of the 10 S&P 500 index sectors based on annual returns from 1990 to 2014. The returns fall roughly into three groups. The first group of six sectors has returns between 10% and 12%; the second group of three sectors has returns between 8% and 9%; and the last group is the telecommunication sector, with the lowest return of 5.7%. The volatilities have a wider range, with financials and technology having the highest risks, and consumer staples and energy having the lowest risk.

For each pair, we calculate the difference of their geometric and pairwise volatilities of their return differences based on historical returns. These measures serve as the basis for calculating volatility and return effects of each pair. Table 7.4 lists the return differences between individual pairs. The element in the ith row and the jth column is $(g_i - g_j)$. Note that the matrix is antisymmetric. Also, since the health care sector has the highest

TABLE 7.3 Annual Geometric Return Statistics of 10 S&P 500 Sectors from 1990 to 2014

	Return (%)	Volatility (%)	Sharpe Ratio
Consumer staples (CSS)	11.31	14.49	0.56
Consumer discretionary (CSD)	10.33	22.26	0.32
Energy (ENE)	10.66	16.80	0.45
Financials (FIN)	8.67	25.39	0.22
Health care (HLH)	12.14	22.68	0.40
Industrials (IND)	10.06	18.84	0.37
Information technology (TEC)	10.95	32.26	0.24
Materials (MAT)	8.14	19.09	0.26
Telecommunication services (TEL)	5.60	22.83	0.11
Utilities (UTL)	8.24	20.63	0.25

TABLE 7.4   Return Differences between Individual Pairs of Two Sectors

	CSS (%)	CSD (%)	ENE (%)	FIN (%)	HLH (%)	IND (%)	TEC (%)	MAT (%)	TEL (%)	UTL (%)
CSS	0.0	1.0	0.7	2.6	−0.8	1.3	0.4	3.2	5.7	3.1
CSD	−1.0	0.0	−0.3	1.7	−1.8	0.3	−0.6	2.2	4.7	2.1
ENE	−0.7	0.3	0.0	2.0	−1.5	0.6	−0.3	2.5	5.1	2.4
FIN	−2.6	−1.7	−2.0	0.0	−3.5	−1.4	−2.3	0.5	3.1	0.4
HLH	0.8	1.8	1.5	3.5	0.0	2.1	1.2	4.0	6.5	3.9
IND	−1.3	−0.3	−0.6	1.4	−2.1	0.0	−0.9	1.9	4.5	1.8
TEC	−0.4	0.6	0.3%	2.3	−1.2	0.9	0.0	2.8	5.3	2.7
MAT	−3.2	−2.2	−2.5	−0.5	−4.0	−1.9	−2.8	0.0	2.5	−0.1
TEL	−5.7	−4.7	−5.1	−3.1	−6.5	−4.5	−5.3	−2.5	0.0	−2.6
UTL	−3.1	−2.1	−2.4	−0.4	−3.9	−1.8	−2.7	0.1	2.6	0.0

return, all the returns in its row are positive. In contrast, all the returns in the row of the telecommunication sector are negative.

Table 7.5 displays volatility of return differences between individual pairs. The element in the ith row and the jth column is σ_{i-j}. Note that the matrix is symmetric, and all off-diagonal elements are positive. Since the technology sector has the highest return volatility, many pairs involving it also have high volatility. Many pairs involving consumer staples have low volatility. However, it is the industrial sector that has the lowest pairwise return volatility overall, the reason being that it has high correlations with many other sectors.

In Table 7.6, we display the combined effect in terms of the t-statistic of all pairs in a matrix form. The element in the ith row and the jth column is simply $\sqrt{24}\left(g_i - g_j\right)/\sigma_{i-j}$, since there are 25 years in the sample.

TABLE 7.5   Pairwise Volatility of Return Differences between Individual Pairs

	CSS (%)	CSD (%)	ENE (%)	FIN (%)	HLH (%)	IND (%)	TEC (%)	MAT (%)	TEL (%)	UTL (%)
CSS	0.0	20.0	16.5	17.9	11.6	14.7	31.6	19.5	21.1	16.0
CSD	20.0	0.0	21.9	18.1	22.8	12.8	22.1	15.7	15.8	25.0
ENE	16.5	21.9	0.0	21.0	22.7	12.5	29.1	15.2	20.9	16.9
FIN	17.9	18.1	21.0	0.0	19.0	13.6	31.2	20.4	23.3	20.8
HLH	11.6	22.8	22.7	19.0	0.0	18.8	31.6	25.8	24.6	17.9
IND	14.7	12.8	12.5	13.6	18.8	0.0	24.4	12.0	17.2	17.8
TEC	31.6	22.1	29.1	31.2	31.6	24.4	0.0	25.0	22.9	34.4
MAT	19.5	15.7	15.2	20.4	25.8	12.0	25.0	0.0	21.3	23.9
TEL	21.1	15.8	20.9	23.3	24.6	17.2	22.9	21.3	0.0	25.0
UTL	16.0	25.0	16.9	20.8	17.9	17.8	34.4	23.9	25.0	0.0

TABLE 7.6   Pairwise t-Statistics $\tilde{t}_{i-j}$ of Return Differences between Distinct Sector Pairs

	CSS	CSD	ENE	FIN	HLH	IND	TEC	MAT	TEL	UTL
CSS		0.24	0.20	0.73	-0.35	0.42	0.06	0.80	**1.33**	0.94
CSD	-0.24		-0.07	0.45	-0.39	0.10	-0.14	0.68	**1.47**	0.41
ENE	-0.20	0.07		0.46	-0.32	0.23	-0.05	0.81	**1.19**	0.70
FIN	-0.73	-0.45	-0.46		-0.90	-0.50	-0.36	0.13	0.65	0.10
HLH	0.35	0.39	0.32	0.90		0.54	0.19	0.76	**1.30**	**1.07**
IND	-0.42	-0.10	-0.23	0.50	-0.54		-0.18	0.78	**1.27**	0.50
TEC	-0.06	0.14	0.05	0.36	-0.19	0.18		0.55	**1.14**	0.39
MAT	-0.80	-0.68	-0.81	-0.13	-0.76	-0.78	-0.55		0.58	-0.02
TEL	**-1.33**	**-1.47**	**-1.19**	-0.65	**-1.30**	**-1.27**	**-1.14**	-0.58		-0.52
UTL	-0.94	-0.41	-0.70	-0.10	**-1.07**	-0.50	-0.39	0.02	0.52	

Note: Those above 1 or below −1 (in bold) indicate potential negative rebalancing alpha for the pair; other pairs have *t*-statistics between −1 and 1, indicating positive rebalancing alpha.

If the *t*-statistic is between −1 and 1, then the rebalancing alpha is positive for the pair. If the *t*-statistic is greater than 1 or less than −1, the rebalancing alpha is negative for the pair. Even though the matrix is antisymmetric, only the magnitude of the matrix elements is relevant to rebalancing alpha.

Out of 45 distinct pairs, 7 have negative rebalancing alpha, of which 6 involve the telecommunication sector, since its return is much lower than that of other sectors. The other pair is between the utilities and the health care sector. The remaining 38 sectors would have positive alpha. Note that two sectors, financials and materials, have positive rebalancing alpha when paired with all the other nine sectors.

In addition to the small proportion of pairs with negative rebalancing alpha, the magnitudes of those negative alphas are also relatively small compared with the magnitudes of positive alphas. This can be seen from the fact that the *t*-statistics of negative pairs are close to either 1 or −1, but the *t*-statistics of many positive pairs are rather small in terms of their magnitudes. For a long-only portfolio composed of these 10 sectors, its rebalancing alpha is a weighted sum of these pairwise rebalancing alphas. One would suspect that it is highly likely that the rebalancing alpha of a well-diversified portfolio will be positive.

We shall not evaluate the accuracy of analytic approximations of volatility and return effects and rebalancing alpha of 45 individual pairs. Instead, we evaluate the accuracy of approximations for a multitude of portfolios. An obvious choice is the equal-weighted portfolio. We display the results in Table 7.7. First, the average of geometric returns is 9.61%, and both FW and BH portfolios' returns are higher, with the FW portfolio returns at 10.53% and the BH portfolio returns at 9.94%. This is a positive rebalancing alpha of 0.58%. The analytical approximation yields a volatility effect of 1.03% versus the actual value of 0.92% (=10.53% to 9.61%). The approximation of the return effect is 0.41%, and of the actual value it is 0.33% (=9.94% to 9.61%). For the rebalancing alpha, the estimate is 0.62% with an error of 0.04%.

- Table 7.7 also lists the approximated values for the geometric return and the FW and BH portfolios, which are the sums of the average $\bar{g}$ and the approximated volatility and return effects, respectively.

The example of an equal-weighted portfolio demonstrates that the approximations are reasonably accurate. To evaluate the accuracy for a general long-only portfolio, we randomly simulate 10,000 initial

TABLE 7.7   Actual and Approximated Geometric Returns of Equal-Weighted FW and BH Portfolios and Associated Volatility Effect, Return Effect, and Rebalancing Alpha

	g_{FW}	g_{BH}	$\bar{g}$	e_v	e_r	α
Actual	10.53%	9.94%	9.61%	0.92%	0.33%	0.58%
Approximation	10.64%	10.02%	–	1.03%	0.41%	0.62%
Error	0.11%	0.07%	–	0.11%	0.07%	0.04%

portfolios* and then follow the resulting 10,000 BH and 10,000 FW portfolios over the course of 25 years based on realized sector returns. To check the accuracy of analytic results, we compare the actual rebalancing alpha $\alpha = g_{FW} - g_{BH}$ and the approximated alpha based on Equation 7.24.

Figure 7.7 shows the scatter plots of 10,000 points representing actual and approximated rebalancing alphas for 10,000 portfolios. The dashed line is the line where two are equal, i.e. $\tilde{\alpha} = \alpha$. The vertical distance from each point to this line is the error of the approximation. First, we notice that the errors are systematically positive when the rebalancing alpha is high, while they are systematically negative when the rebalancing alpha is low. This is mainly due to the overestimation of volatility effect when the rebalancing alpha is high, which could be the result of high portfolio volatility. As we indicated before, the estimated volatility effect is vulnerable if portfolio volatility or sector volatilities are high.

Second, we note that the absolute errors are reasonably small despite the observed tilt. Out of 10,000 portfolios, the maximum error is 19 bps, while the average error is 4 bps. A linear regression of approximation against the true alpha has an R-squared of 98%.

Lastly, the rebalancing alphas for 10,000 portfolios are all positive, ranging from 8 to 91 bps. The average rebalancing alpha is 56 bps, which is close to the rebalancing alpha of an equal-weighted portfolio. This extraordinary result is a manifestation of the fact that a majority of pairwise alphas is positive, and none of our randomly simulated portfolios is concentrated in those negative pairs. However, what is not true is that all sector portfolios would have a positive rebalancing alpha. We can easily identify a portfolio with negative rebalancing alpha by choosing in Table 7.6 a pair with the t-statistic above 1. For example, a 50/50 portfolio in consumer

---

* The simulation procedure is analogous to a monkey throwing darts onto a board with 10 equally divided slices. In the first step, we generate 10 numbers based on a random variable with uniform distribution in the interval [0,1]. In the second step, we rescale the 10 numbers as weights so that they sum to 1. Even though the mean weights tend to be equal, 10,000 portfolios exhibit considerable variation.

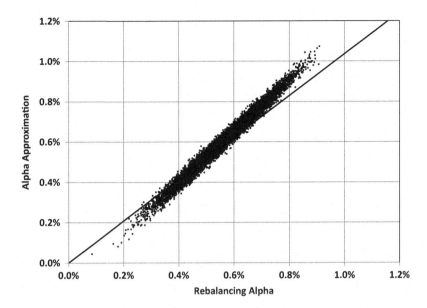

FIGURE 7.7 Scatter plot of rebalancing alphas and their approximations for the 10,000 random simulated portfolios. The solid line is the line when there is no error.

discretionary and telecommunication has a rebalancing alpha of −34 bps. Nevertheless, the probability of negative rebalancing alpha for a diversified sector portfolio is extremely low.

7.2.5 Cross-Sectional Serial Correlations of Sector Portfolios

Why are the sector returns from this period so overwhelmingly favorable to FW portfolios? There are only two possible reasons: high volatility effects and/ or low return effects. An analysis of cross-sectional serial correlations shows that they had a significant impact on lowering the return effects of portfolio rebalancing, resulting in positive rebalancing alpha. We devote a separate section to this analysis for the sector portfolios, which will also be an integral part of the empirical analysis of other markets in the next few chapters.

We study the impact of serial correlations using Equation 6.18, which gives an approximate relationship between serial correlations of all lags and the variance of arithmetic means. For convenience, we copy the equation here as

$$\text{var}(\mu) = \frac{\text{var}(\mathbf{r})}{N}\left[1 + \frac{2}{N}\sum_{k=1}^{N-1}(N-k)\rho(k)\right] \tag{7.30}$$

Equation 7.30 is valid for any portfolios whose sector weights serve as weights in the variance calculation. For brevity, we show the results for one specific portfolio—the equal-weight portfolio.

Figure 7.8 displays the average cross-sectional serial correlations of annual returns over the period from 1990 to 2014 as well as their scaled contribution to the overall variance of arithmetic means. The scaling factor given in Equation 7.30 is $2(N-k)/N$, with k being the lag.

We make several observations. First, the average serial correlation with one lag is positive. This is consistent with a 12-month cross-sectional sector momentum, a part of the equity momentum phenomenon. However, from Lag 2 to Lag 10, the correlations are mostly negative, indicating return reversal at between 2 and 10 years. Beyond 10 years, the correlations seem to alter signs frequently and randomly. Second, the contribution to the variance of returns of each lagged correlation varies. When the lag is short, the contribution is magnified, since there are more sub-periods with short lags. When the lag is long, the contribution is reduced, since there are few sub-periods with long lags. As a result, the correlations with lags beyond 13 years have their contributions scaled down, while the contributions within 12 years are scaled up. Half of 25 is 12.5.

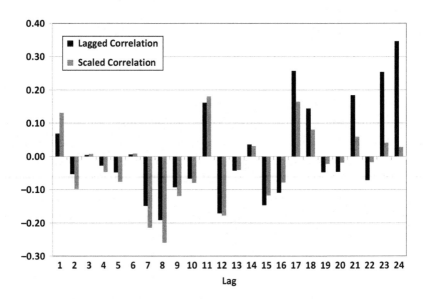

FIGURE 7.8 Average equal-weighted cross-sectional serial correlation of sector returns for different lags and their contribution to the variance of arithmetic means.

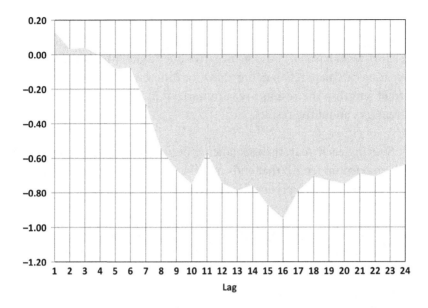

FIGURE 7.9 Cumulative contributions of cross-sectional serial correlations to the variance reduction of arithmetic means.

The cumulative sum of all contributions gives rise to the summation term in Equation 7.30, which measures the overall rise or fall in the variance of arithmetic means caused by cross-sectional serial correlation. In Figure 7.9, we plot the cumulative sum as a function of lags included. We note that the sum is positive up to Lag 4 due to the lingering effect of 12-month momentum. It turns negative at Lag 5 and then declines sharply from Lag 6 to Lag 10, as the correlations and contributions from those lags are significantly negative. As the lag increases further, the cumulative sum settles down and fluctuates around −0.6. From Equation 7.30, this implies a 60% reduction in the variance of arithmetic means relative to a base case in which the sector returns are serially uncorrelated. Another observation from Figure 7.8 is that this level of reduction is achieved when the lag gets to 9 years and beyond. This strongly implies a significant reduction in return effect with a horizon beyond 9 years and perhaps, a high probability of positive rebalancing alpha for those horizons. On the other hand, when the horizon is much shorter, the probability of positive rebalancing will be much smaller.

7.2.6 Rebalancing Alpha of Varying Horizons

The analysis so far applied to the entire period from 1990 to 2014, for which the rebalancing alpha is 58 bps per annum. Of course, rebalancing alpha varies with time horizon and "vintage" year. If asset returns have long-term mean reversion, rebalancing alpha is most likely positive. However, on a short-term basis, it could be negative due to short-term momentum.

We thus calculate rebalancing alpha for different starting years and for different lengths. The results are summarized in Table 7.8. We make several remarks about the results.

- Starting each year, the rebalancing period ranges from two years to a maximum length that ends in 2014. The longest starts in 1990 and ends in 2014, as has been discussed previously, and has a rebalancing alpha of 58 bps.

- For a rebalancing period of two years, in the second column, the rebalancing alpha is rather volatile. The minimum occurs in the 1998–1999 period, giving a rebalancing alpha of −1.37%. On the other hand, the maximum occurs in 1999–2000, giving a rebalancing alpha of 2.85%. The three-year period from 1998 to 2000 coincided with the height of the technology bubble and its subsequent bursting. The sector returns of 1998 and 1999 show a strong momentum effect leading to significant negative rebalancing alpha. The sector returns of 2000 exhibit a strong reversal from those of 1999, leading to significant positive rebalancing alpha.

- The effect of these three years carries over to results covering the horizon of multi-year rebalancing. It is evident in the table that all rebalancing periods ending in 1999 have negative alpha, while all rebalancing periods ending in 2000 have positive alpha.

- More generally, the rebalancing alpha during 1999 and 2000 is so significant that any rebalancing period that includes the year 1999 and 2000 always has positive alpha.

- On the other hand, when 1999 and 2000 are absent from a rebalancing period, the rebalancing alpha tends to be insignificant. We note that the lower portion of the table after the year 2000 contains rebalancing alphas that are insignificant.

- Overall, the analysis for varying horizon points to the impact of returns in 1998, 1999, and 2000 on the rebalancing alpha throughout the entire period.

TABLE 7.8 Rebalancing Alpha of Different Length (Horizontal Axis) and Different Starting Year (Vertical Axis)

	2	3	4	5	6	7	8	9	10	11	12	13	14	15	16	17	18	19	20	21	22	23	24	25
1990	-0.09%	0.40%	0.54%	0.48%	0.39%	0.17%	0.04%	-0.18%	-0.49%	0.52%	0.63%	0.59%	0.56%	0.66%	0.60%	0.65%	0.70%	0.68%	0.72%	0.72%	0.66%	0.68%	0.60%	0.58%
1991	0.03%	0.10%	0.35%	0.30%	0.08%	-0.08%	-0.23%	-0.63%	0.58%	0.65%	0.61%	0.52%	0.62%	0.57%	0.60%	0.75%	0.83%	0.85%	0.82%	0.80%	0.80%	0.71%	0.70%	
1992	-0.23%	0.13%	0.22%	-0.02%	-0.08%	-0.38%	-1.16%	0.63%	0.78%	0.81%	0.67%	0.72%	0.63%	0.64%	0.68%	0.78%	0.79%	0.74%	0.72%	0.75%	0.69%	0.71%		
1993	0.12%	0.14%	-0.18%	-0.21%	-0.65%	-1.49%	0.51%	0.74%	0.78%	0.67%	0.74%	0.62%	0.66%	0.63%	0.67%	0.69%	0.68%	0.63%	0.68%	0.61%	0.63%			
1994	-0.14%	-0.45%	-0.52%	-0.94%	-1.49%	0.46%	0.69%	0.65%	0.61%	0.74%	0.61%	0.67%	0.65%	0.63%	0.68%	0.74%	0.62%	0.66%	0.58%	0.58%				
1995	-0.19%	-0.40%	-0.71%	-1.16%	0.85%	1.00%	0.90%	0.83%	0.90%	0.74%	0.76%	0.78%	0.76%	0.85%	0.82%	0.74%	0.78%	0.70%	0.70%					
1996	-0.05%	-0.41%	-1.23%	1.34%	1.37%	1.19%	1.03%	1.04%	0.86%	0.85%	0.81%	0.79%	0.86%	0.80%	0.73%	0.78%	0.70%	0.72%						
1997	-0.45%	-1.00%	1.94%	1.79%	1.51%	1.35%	1.29%	1.07%	1.02%	0.94%	0.85%	0.95%	0.88%	0.77%	0.82%	0.76%	0.77%							
1998	-1.37%	2.67%	2.27%	1.80%	1.53%	1.39%	1.12%	1.05%	0.87%	0.80%	0.88%	0.79%	0.70%	0.78%	0.72%	0.76%								
1999	2.85%	2.00%	1.09%	0.87%	0.68%	0.32%	0.29%	0.06%	0.17%	0.35%	0.26%	0.24%	0.40%	0.38%	0.50%									
2000	0.44%	-0.40%	-0.10%	-0.12%	-0.45%	-0.38%	-0.51%	-0.49%	-0.10%	-0.10%	-0.17%	0.05%	0.07%	0.15%										
2001	-0.41%	-0.26%	-0.19%	-0.26%	-0.25%	-0.40%	-0.24%	-0.06%	-0.16%	-0.13%	0.03%	0.04%	0.20%											
2002	0.06%	0.09%	-0.17%	-0.41%	-0.42%	-0.29%	-0.03%	-0.08%	-0.11%	0.09%	0.15%	0.27%												
2003	0.11%	-0.13%	-0.14%	-0.59%	-0.06%	0.03%	0.01%	-0.09%	0.13%	0.21%	0.29%													
2004	-0.25%	-0.32%	-0.64%	0.09%	-0.06%	-0.06%	-0.20%	0.05%	0.17%	0.26%														
2005	-0.01%	-0.34%	-0.38%	0.18%	0.11%	-0.07%	0.16%	0.23%	0.28%															
2006	-0.05%	-0.23%	0.21%	-0.06%	-0.02%	0.16%	0.23%	0.25%																
2007	-0.28%	0.15%	0.18%	0.15%	0.14%	0.16%	0.14%																	
2008	0.42%	0.34%	0.07%	0.18%	-0.02%	0.01%																		
2009	-0.10%	0.20%	0.16%	0.15%	0.06%																			
2010	0.14%	0.21%	0.10%	0.11%																				
2011	0.23%	0.13%	-0.01%																					
2012	-0.19%	-0.16%																						
2013	-0.04%																							

PROBLEMS

7.1. Derive Equation 7.10.

7.2. Derive Equations 7.17 through 7.19.

For a two-asset long-short portfolio, in which $w_1 < 0$, $w_2 > 0$, and $w_1 + w_2 = 1$, prove that g_2 is always between g_1 and $\bar{g}$.

7.3. Show that when $\left[\sigma_{i-j}^2 - (N-1)(g_i - g_j)^2 \right]$ is the same for all distinct pairs, the equally weighted portfolio has an extreme value of rebalancing alpha.

Asset Allocation Portfolios

I N THIS CHAPTER AND the next, we empirically examine rebalancing alpha of different types of portfolio that investors encounter in practice. We study asset allocation portfolios in Chapter 8 and asset class portfolios in stocks, bonds, and commodities in Chapter 9. The examination will follow along the lines we have used in Chapter 7, wherein we examined sector portfolios based on the sectors in the S&P 500 index. The objectives are two-fold. First, we objectively examine rebalancing alphas of different portfolios. Second, we use analytical results obtained in the previous chapters to explain and understand the nature and the magnitude of rebalancing alpha. Together, these analyses will help us understand certain characteristics of return distributions of asset classes.

8.1 TRADITIONAL 60/40 PORTFOLIOS

Traditional asset allocation portfolios typically invest 60% of capital in stocks and 40% in bonds. There are no solid theoretical reasons why many investors have adopted this asset allocation scheme, which has been dubbed a *balanced* portfolio because of its seemingly balanced allocation of capital. However, from a risk perspective, the 60/40 portfolio is a rather concentrated portfolio in its equity risk allocation. The primary practical reason for its popularity among institutional investors seems to be that over time, it has been capable of delivering an annual return of about 7%, which is required by these investors. The return of the 60/40 portfolio would be heavily dependent on the return of the equity market because of the risk concentration. For this reason, the delivery of 7% return has been highly unstable and may be unreliable in the future.

In terms of rebalancing alpha, a 60/40 portfolio would be very similar to the 50/50 portfolio we briefly discussed in Chapter 3 (Table 3.10). There, we used just two asset classes, and the rebalancing alpha was close to zero from 1970 to 2014. Here, we provide another example with more asset classes, which is more realistic in representing asset allocation by institutional investors.

Table 8.1 displays the annualized return and standard deviation of eight asset classes based on annual returns from 1988 to 2014. There are four equity asset classes and four fixed income asset classes. However, the high yield bonds are dominated by equity risks and they are highly correlated equity market. Therefore, we will classify high yield bonds as an equity asset class when we consider investments in a 60/40 portfolio. The five equity asset classes have relatively high risk, ranging from 16.26% in high yield to 34.91% in emerging market equity. They also have high returns, except for the international stocks, which returned only 3.69% on an annual basis. The three fixed income asset classes have lower returns, but their risks are much lower. As a result, the Sharpe ratios of the three fixed income asset classes are much higher than the Sharpe ratios of equity asset classes. The Sharpe ratio of international stocks is the lowest.

Table 8.2 shows the pairwise differences among all possible pairs. Two patterns are clear: one, stocks in general have outperformed bonds; two, international stocks have underperformed all other asset classes.

Table 8.3 shows the volatility of return differences among all pairs. The volatility is low among fixed income pairs and high for all other pairs, whether it is between two equity asset classes or between one equity and one fixed income asset class. Finally, any pair involving emerging market (EM) stocks has extremely high volatility of return differences.

TABLE 8.1 Return Statistics of Asset Classes Based on Annual Returns from 1988 to 2014 used in the 60/40 Portfolio

	U.S. Large Cap	U.S. Small Cap	Int'l Stocks	EM Stocks	High Yield	U.S. Treasury	WGBI ex U.S.	U.S. Corp.
Return	10.79%	10.51%	3.69%	11.57%	8.60%	6.53%	6.47%	7.53%
Risk	18.16%	19.46%	19.41%	34.91%	16.26%	5.97%	4.60%	6.50%
Sharpe	0.41	0.36	0.01	0.23	0.32	0.52	0.66	0.63

TABLE 8.2   Pairwise Return Difference among Asset Classes

	U.S. Large Cap (%)	U.S. Small Cap (%)	Int'l Stocks (%)	EM Stocks (%)	High Yield (%)	U.S. Treasury (%)	WGBI ex U.S. (%)	U.S. Corp. (%)
U.S. Large Cap	0.00	0.28	7.10	−0.77	2.20	4.26	4.32	3.26
U.S. Small Cap	−0.28	0.00	6.82	−1.05	1.92	3.98	4.04	2.98
Int'l Stocks	−7.10	−6.82	0.00	−7.87	−4.90	−2.84	−2.78	−3.84
EM Stocks	0.77	1.05	7.87	0.00	2.97	5.04	5.10	4.03
High Yield	−2.20	−1.92	4.90	−2.97	0.00	2.07	2.13	1.06
U.S. Treasury	−4.26	−3.98	2.84	−5.04	−2.07	0.00	0.06	−1.00
WGBI ex U.S.	−4.32	−4.04	2.78	−5.10	−2.13	−0.06	0.00	−1.06
U.S. Corp.	−3.26	−2.98	3.84	−4.03	−1.06	1.00	1.06	0.00

TABLE 8.3   Pairwise Volatility of Return Differences among Asset Classes

	U.S. Large Cap (%)	U.S. Small Cap (%)	Int'l Stocks (%)	EM Stocks (%)	High Yield (%)	U.S. Treasury (%)	WGBI ex U.S. (%)	U.S. Corp. (%)
U.S. Large Cap	0.00	10.50	12.95	30.26	14.44	20.06	18.05	16.50
U.S. Small Cap	10.50	0.00	14.66	27.37	12.34	21.75	19.80	18.03
Int'l Stocks	12.95	14.66	0.00	25.90	18.55	22.43	19.84	19.69
EM Stocks	30.26	27.37	25.90	0.00	27.39	36.99	35.92	33.46
High Yield	14.44	12.34	18.55	27.39	0.00	18.42	16.58	13.10
U.S. Treasury	20.06	21.75	22.43	36.99	18.42	0.00	4.44	6.31
WGBI ex U.S.	18.05	19.80	19.84	35.92	16.58	4.44	0.00	5.79
U.S. Corp.	16.50	18.03	19.69	33.46	13.10	6.31	5.79	0.00

Table 8.4 shows the pairwise t-statistics of all pairs of asset classes. Recall that if the absolute value of the t-statistic is less than 1, the rebalancing alpha of a long-only portfolio involving the pair is positive. On the other hand, if the absolute value of the t-statistic is greater than 1, the rebalancing alpha of a long-only portfolio involving the pair is negative. Out of 28 unique pairs, there are some pairs whose t-statistics are very close to 1 in terms of absolute value. There are five pairs with negative rebalancing alpha, four of which involve international stocks. This is because the return is so much lower than that of other asset classes. The remaining pair is between the U.S. large cap stocks and World Government Bond Index (WGBI) ex U.S. government bonds.

We choose two 60/40 portfolios with portfolio weights listed in Table 8.5. We note that Portfolio 1 is a reasonable strategic 60/40 asset allocation portfolio. First, note that we count high yield as an equity asset

TABLE 8.4   Pairwise *t*-Statistics among Different Pairs

	U.S. Large Cap	U.S. Small Cap	Int'l Stocks	EM Stocks	High Yield	U.S. Treasury	WGBI ex U.S.	U.S. Corp.
U.S. Large Cap	0.00	0.13	**2.80**	−0.13	0.78	**1.08**	**1.22**	**1.01**
U.S. Small Cap	−0.13	0.00	**2.37**	−0.20	0.79	0.93	**1.04**	0.84
Int'l Stocks	**−2.80**	**−2.37**	0.00	**−1.55**	**−1.35**	−0.64	−0.71	−0.99
EM Stocks	0.13	0.20	**1.55**	0.00	0.55	0.69	0.72	0.61
High Yield	−0.78	−0.79	**1.35**	−0.55	0.00	0.57	0.65	0.41
U.S. Treasury	**−1.08**	−0.93	0.64	−0.69	−0.57	0.00	0.07	−0.81
WGBI ex U.S.	**−1.22**	**−1.04**	0.71	−0.72	−0.65	−0.07	0.00	−0.94
U.S. Corp.	**−1.01**	−0.84	0.99	−0.61	−0.41	0.81	0.94	0.00

Those with absolute value above 1 are in bold face.

TABLE 8.5   Portfolio Weights of Two 60/40 Portfolios

	U.S. Large Cap (%)	U.S. Small Cap (%)	Int'l Stocks (%)	EM Stocks (%)	High Yield (%)	U.S. Treasury (%)	WGBI ex U.S. (%)	U.S. Corp. (%)
Portfolio 1	20	10	20	5	5	15	15	10
Portfolio 2	30	15	0	10	5	15	15	10

class. Second, it is typical for institutional investors to allocate more to large cap stock (both U.S. and international) than to small cap and emerging market stocks. Within fixed income assets, it is also true that allocation to government bonds is often higher than allocation to corporate bonds. Portfolio 2 is perhaps less typical due to the removal of international stocks. The reason for studying this portfolio is that most pairs with negative rebalancing alpha in Table 8.4 are related to international stocks due to the low cumulative return from this class. By not allocating to it, we would certainly increase rebalancing alpha. The question is by how much.

Based on these portfolio weights, we calculate returns of the fixed-weight (FW) and buy-and-hold (BH) portfolios using actual annual returns. We also derive approximated volatility and return effects using analytic results derived in previous chapters. Both actual results and approximations of Portfolio 1 are displayed in Table 8.6. The return of the BH portfolio is higher than that of the FW portfolio by 8 basis points (bps), resulting in a negative rebalancing alpha. Both volatility and return effects are positive, at 0.73% and 0.81%, respectively. The approximated volatility and return effects are 0.76% and 0.93%, respectively. This indicated a

TABLE 8.6 Returns of FW and BH Portfolios Based on 60/40 Asset allocation Portfolio 1 and Volatility, Return Effects, and Rebalancing Alpha

	g_{FW}	g_{BH}	$\bar{g}$	e_v	e_r	α
Actual	8.39%	8.47%	7.66%	0.73%	0.81%	−0.08%
Approximation				0.76%	0.93%	−0.17%

TABLE 8.7 Returns of FW and BH Portfolios Based on 60/40 Asset allocation Portfolio 2 and Volatility, Return Effects, and Rebalancing Alpha

	g_{FW}	g_{BH}	$\bar{g}$	e_v	e_r	α
Actual	9.90%	9.55%	9.10%	0.80%	0.45%	0.35%
Approximation				0.88%	0.53%	0.35%

rebalancing alpha of −17 bps. The error for the return effect accounts for the majority of the error in rebalancing alpha.

The results for Portfolio 2 are displayed in Table 8.7. As expected, and with hindsight, by not allocating to international stocks, we bring the rebalancing alpha of this portfolio to +35 bps. Approximations to both the volatility and return effects are higher than the actual value by 8 bps. As a result, the approximation of the rebalancing alpha is quite accurate.

- The results from the two portfolios illustrate that the rebalancing alpha is highly dependent on portfolio allocation and return differences between different asset classes. We shall see that the rebalancing alpha is also highly dependent on specific periods in time.

The rest of the analysis will be focused on Portfolio 1, which is more diversified than Portfolio 2. Figure 8.1 displays the average serial cross-correlations of the eight asset classes as well as their scaled contribution over the period from 1988 to 2014. From one to three years, the correlations are quite positive, indicating cross-sectional momentum. From 4 to 13 years, the correlations are on average negative, indicating mild mean reversion. From 14 to 21 years, the correlations turn positive again. The rest of the correlations from 22 years on are quite random. Overall, the scaled contributions from serial cross-sectional correlations are positive. This would lead to a barrier for positive rebalancing alpha for this period.

This fact is reflected in Figure 8.2, which shows the cumulative contribution from cross-sectional serial correlations to the variance of arithmetic means. Recall that the return effect is proportional to the cross-sectional variance of geometric means, which is highly dependent

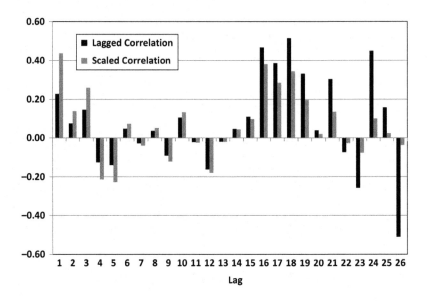

FIGURE 8.1 Average cross-sectional serial correlation of asset returns with respect to weights of Portfolio 1 for different lags and their scaled contribution to the variance of arithmetic means.

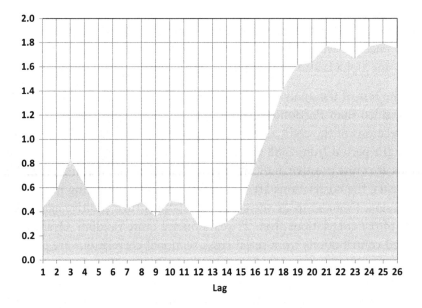

FIGURE 8.2 Cumulative contributions of cross-sectional serial correlations to the variance reduction of arithmetic means.

on the cross-sectional variance of arithmetic means. Figure 8.2 shows that cross-sectional serial correlations contribute positively to the variance of arithmetic means over this period. It is not surprising that the rebalancing alpha is slightly negative for Portfolio 1.

Shown in Figures 8.1 and 8.2, the cross-sectional serial correlations up to 13 years, on average, make only a slightly positive contribution to the variance of arithmetic means. This indicates that rebalancing alpha with an investment horizon of 13 years could potentially be positive. Hence, we compute returns of FW and BH portfolios with a 13-year investment horizon, and the results are shown in Figure 8.3. The starting year of the investment period is indicated on the x-axis. For instance, the year 1988 indicates the period from 1988 to 2000, and the year 2002 indicates the period from 2002 to 2014. Indeed, the rebalancing alpha is generally positive except for the first three cases, where the starting year was 1988, 1989, and 1990. These results show that the rebalancing alpha of asset allocation portfolios is sensitive to investment horizon.

- In addition, we note that the overall return has declined in recent periods. For example, returns from 1988 to 2000 were close to 11% for both portfolios, while returns from 2002 to 2014 were only about 6%.

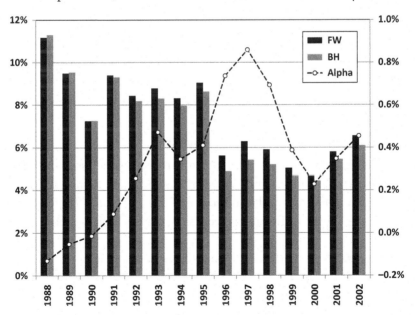

FIGURE 8.3   Returns of FW and BH portfolios and rebalancing alphas of investment horizon of 13 years.

8.2 RISK PARITY PORTFOLIOS

Risk parity is a relatively new investment approach that belongs to the broad category of risk-based investing. Risk parity originated as an investment approach in asset allocation, and it has been extended to other investment areas such as equity, fixed income, commodities, and other asset classes or strategies. The latter might include multi-strategy hedge funds and multi-factor quantitative factor models.

In this section, we study rebalancing alpha of asset allocation portfolios based on a risk parity approach, because these portfolios are often leveraged investments with short positions in cash or risk-free rate. The risk parity approach dictates that a portfolio must have larger exposure to low-risk assets and smaller exposure to high-risk assets. For example, if bonds have a risk of 5% and stocks have a risk of 15%, a simple (or naïve) risk parity without leverage would invest 75% in bonds and 25% in stocks. While this 25/75 (in stock/bond) portfolio is much more diversified than the traditional 60/40 portfolio, the portfolio total risk and expected return would be low. To provide adequate returns to investors, one would consider leveraging the portfolio to 200% exposure with 50% in stocks and 150% in bonds. This can be achieved easily by investing in the futures market. This leverage is not free, and in fact, it is financed by a short position in cash. Therefore, the portfolio weight should be considered as −100/50/150 in cash/stock/bond assets rather than 50/150 in stock/bond.

In this section, we examine rebalancing alpha for leveraged risk parity portfolios. We proceed in two steps. The first is to construct and study an unlevered risk parity portfolio. In the second step, we analyze portfolios with leverage.

8.2.1 Risk Parity Portfolio without Leverage

We construct risk parity portfolios with three asset classes introduced previously in Chapter 2. They are stocks, bonds, and commodities, and their return statistics (similar to those in Table 2.1) from 1970 to 2014 are displayed in Table 8.8. The unlevered risk parity portfolio has no zero cash

TABLE 8.8   Return Statistics of Asset Classes Based on Annual Returns from 1970 to 2014

	Cash	Bonds	Stocks	Commodities
g	5.16%	7.58%	10.48%	8.03%
σ	3.38%	7.60%	17.24%	24.66%
Sharpe		0.32	0.31	0.12

TABLE 8.9 Portfolio Weight of Unlevered Risk Parity Asset allocation Portfolio

	Cash (%)	Bonds (%)	Stocks (%)	Commodities (%)
Unlevered Risk Parity	0	58	25	17

weight. The correlations among the rest of the three asset classes are minimal. Therefore, the risk parity weights are roughly inversely proportional to return volatilities, listed in Table 8.9, with 58% in bonds, 25% in stocks, and 17% in commodities.

Table 8.10 lists the pairwise cumulative return differences between four asset classes. Even though cash is not part of the unlevered risk parity portfolio, it will be part of levered ones (with a short position). Therefore, we include it as part of the analysis. Table 8.11 displays the volatilities of pairwise return difference. Since commodity returns are the most volatile, any pair involving commodities has high return volatility. The cash/bond pair has the lowest volatility in return differences. Table 8.12 shows the pairwise t-statistics. Two pairs, cash/bond and cash/stock, have the highest t-statistic in magnitude. The stock/bond pair has a t-statistic slightly above 1 in magnitude. In a long-only portfolio, these three pairs would generate negative rebalancing alpha, while the other three pairs would generate positive rebalancing alpha. It is fair to expect that any long-only portfolio would generate positive rebalancing alpha.

TABLE 8.10 Pairwise Return Difference among Four Asset Classes

	Cash (%)	Bonds (%)	Stocks (%)	Commodities (%)
Cash	0.00	−2.42	−5.32	−2.87
Bonds	2.42	0.00	−2.90	−0.45
Stocks	5.32	2.90	0.00	2.45
Commodities	2.87	0.45	−2.45	0.00

TABLE 8.11 Pairwise Volatility of Return Differences among Four Asset Classes

	Cash (%)	Bonds (%)	Stocks (%)	Commodities (%)
Cash	0.00	7.41	17.63	24.46
Bonds	7.41	0.00	18.93	27.33
Stocks	17.63	18.93	0.00	31.51
Commodities	24.46	27.33	31.51	0.00

TABLE 8.12   Pairwise t-Statistics among Different Pairs

	Cash	Bonds	Stocks	Commodities
Cash	0.00	−2.17	−2.00	−0.78
Bonds	2.17	0.00	−1.02	−0.11
Stocks	2.00	1.02	0.00	0.52
Commodities	0.78	0.11	−0.52	0.00

TABLE 8.13   Returns of FW and BH Portfolios Based on Risk Parity Asset allocation in Table 8.9 and Volatility, Return Effects, and Rebalancing Alpha

	g_{FW}	g_{BH}	$\bar{g}$	e_v	e_r	α
Actual	9.20%	8.73%	8.38%	0.82%	0.35%	0.47%
Approximation				0.84%	0.33%	0.51%

We calculate returns of the FW and BH portfolios using actual annual returns, and derive approximated volatility and return effects. Both actual results and approximations of a risk parity unlevered portfolio are displayed in Table 8.13. The return of the FW portfolio is higher than that of the BH portfolio by 47 bps, resulting in a positive rebalancing alpha. Both volatility and return effects are positive, at 0.82% and 0.35%, respectively. The approximated volatility and return effects are 0.84% and 0.33%, respectively. This indicates a rebalancing alpha of 51 bps.

Figure 8.4 displays the average serial cross-correlations of the three asset classes as well as their scaled contribution over the period from 1970 to 2014. It appears that there is no strong pattern, and positive and negative correlations are balanced. Figure 8.5 shows the cumulative contribution of cross-sectional serial correlations to the variance of arithmetic means over this period. It grows quite positively from Lag 14 to Lag 34. Still, the rebalancing alpha turns out to be positive.

Again, we compute returns of FW and BH portfolios with a 13-year investment horizon, and the results are shown in Figure 8.6. The starting year of the investment period is indicated on the horizontal axis. Out of 33 different and overlapping periods, 26 have positive rebalancing alpha. These results show that the rebalancing alpha of asset allocation portfolios is sensitive to investment horizon. All periods with negative rebalancing alpha started in the 1980s. This is probably due to the fact that a bull market in equities started in the 1980s and lasted until 1999. Consistently with the previous case, the return of this portfolio has also been declining in recent years.

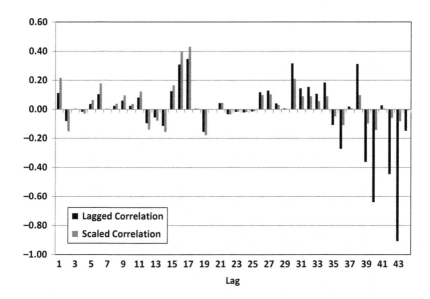

FIGURE 8.4 Average cross-sectional serial correlation of asset returns with respect to weights of unlevered risk parity portfolio for different lags and their scaled contribution to the variance of arithmetic means.

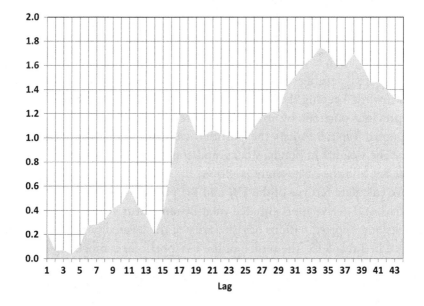

FIGURE 8.5 Cumulative contributions of cross-sectional serial correlations to the variance reduction of arithmetic means.

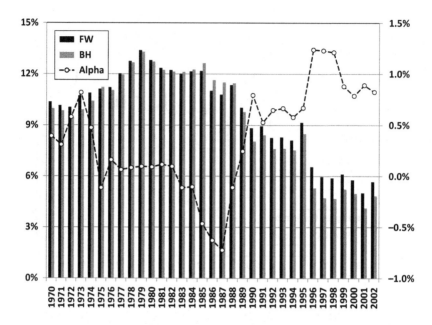

FIGURE 8.6   Returns of FW and BH risk parity portfolios and rebalancing alphas of investment horizon of 13 years.

8.2.2 Risk Parity Portfolio with Leverage

The unlevered risk parity portfolio has a good Sharpe ratio due to risk diversification. However, its risk and expected return are low due to high capital allocation to low-risk assets of bonds. To increase expected return, one can apply leverage to the portfolio, thus increasing risk and expected return while keeping the Sharpe ratio the same. In this section, we lever the previous portfolio by applying 200% leverage and study the rebalancing alpha. Table 8.14 lists the portfolio weights in the asset classes and cash. The weights in bonds, stocks, and commodities are double those in Table 8.9, and the cash weight is −100%.

We calculate returns of the FW and BH portfolios using actual annual returns, and derive approximated volatility and return effects. Both actual results and approximations of risk parity of the levered portfolio are displayed in Table 8.15. The return of the FW portfolio is much higher than

TABLE 8.14   Portfolio Weight of Levered Risk Parity Asset allocation Portfolio

	Cash (%)	Bonds (%)	Stocks (%)	Commodities (%)
Levered Risk Parity	−100	116	50	34

TABLE 8.15 Returns of FW and BH Portfolios Based on Leveraged Risk Parity Asset allocation in Table 8.14 and Volatility, Return Effects, and Rebalancing Alpha

	g_{FW}	g_{BH}	$\overline{g}$	e_v	e_r	α
Actual	12.85%	10.13%	11.61%	1.24%	−1.48%	2.72%
Approximation				1.24%	−3.51%	4.75%

that of the BH portfolio in the levered case. The volatility effect is a positive 1.24%, while the return effect is a negative 1.48%, resulting in a rebalancing alpha of 2.72%. The BH approach results in declining leverage over time, limiting its return potential.

The approximated volatility effect is 1.24%, while the approximated return effect shows a significant error compared with the actual result. As indicated in Chapter 6, for a long-short portfolio, return effect approximation is incorrect with formulae that are derived with long-only weights. In Table 8.16, we recalculated the return effect approximation using modified Equation 6.27. The new value is −1.00%, resulting in less error.

Figure 8.7 displays the scaled average serial cross-"correlations" of the four asset classes (including cash) over the period from 1970 to 2014. Since one of the weights is negative, these correlations are not correlations in the usual sense. It appears that there is a strong pattern of negative correlations. Figure 8.8 shows the cumulative contribution of cross-sectional serial correlations to the variance of arithmetic means over this period. It grows quite negative consistently. Perhaps it is not surprising that the rebalancing alpha turns out to be very positive.

We compute returns of FW and BH portfolios with a 13-year investment horizon, and the results are shown in Figure 8.9. The starting year of the investment period is indicated on the horizontal axis. Out of 33 different and overlapping periods, 32 have positive rebalancing alpha. It is essential that we rebalance levered risk parity portfolios. Consistently with previous cases, the return of this portfolio has also been declining in recent years. However, the FW portfolio has maintained returns above 8%.

TABLE 8.16 Returns of FW and BH Portfolios for Leveraged Risk Parity Portfolio and Volatility, Return Effects, and Rebalancing Alpha

	g_{FW}	g_{BH}	$\overline{g}$	e_v	$e_{r(LS)}$	α
Actual	12.85%	10.13%	11.61%	1.24%	−1.48%	2.72%
Approximation				1.24%	−1.00%	2.24%

Note: The Return Effect Approximation Is Calculated with the Long-Short Formula (Equation 6.27).

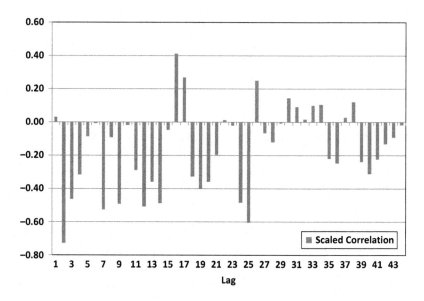

FIGURE 8.7   Average scaled cross-sectional serial "correlation" of asset returns with respect to weights of levered risk parity portfolio for different lags.

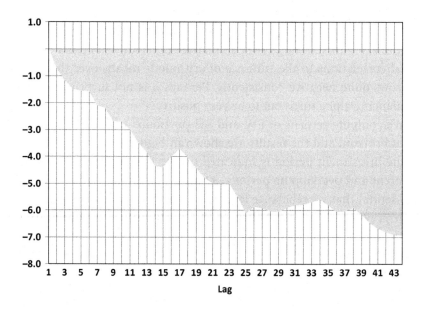

FIGURE 8.8   Cumulative contributions of cross-sectional serial correlations for the levered risk parity portfolio.

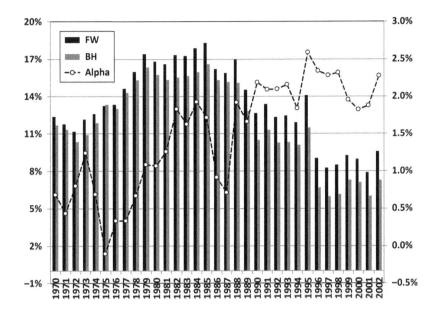

FIGURE 8.9   Returns of FW and BH levered risk parity portfolios and rebalancing alphas of investment horizon of 13 years.

Asset Class Portfolios

IN THIS CHAPTER, WE empirically examine rebalancing alpha of asset class portfolios in stocks, bonds, and commodities. The examination will follow along the lines we have used in Chapter 8. First, we objectively examine rebalancing alphas of different portfolios. Second, we use analytical results obtained in the previous chapters to explain and understand the nature and the magnitude of rebalancing alpha.

9.1 STOCK PORTFOLIOS

The two fundamental dimensions of global equity markets are country and sector. We have analyzed S&P 500 sector portfolios in Chapter 7. In this chapter, we examine stock portfolios consisting of individual countries.

The country returns we use are Morgan Stanley Capital International (MSCI) country index returns in U.S. dollar terms, or unhedged returns. There are 23 developed countries in the MSCI world index. For the purpose of illustration, we shall select 10 countries for our analysis. They represent major economies of the world and also major countries in terms of index weights. Their return series go back to 1970. Other countries in the MSCI world index often have a shorter return history.

Table 9.1 displays annualized return, standard deviation, and Sharpe ratio of the 10 countries, based on annual returns from 1970 to 2016. They are the United States, Canada, Japan, Australia, Hong Kong, Singapore, the United Kingdom, Germany, France, and Italy. The annualized geometric returns are mostly around 9–10% except for Italy and Hong Kong. Italy's return is slightly below 5%, while Hong Kong's return is above 14%. The risk or standard deviation of returns is more different across countries.

TABLE 9.1 Return Statistics of 10 Equity Markets Based on Annual Returns from 1970 to 2016

	United States	Canada	Japan	Australia	Hong Kong	Singapore	United Kingdom	Germany	France	Italy
Return	10.01%	9.27%	9.09%	8.83%	14.27%	10.37%	9.55%	9.56%	9.68%	4.94%
Risk	17.36%	21.68%	32.02%	26.15%	44.12%	44.91%	27.34%	29.02%	27.19%	34.47%
Sharpe	0.29	0.20	0.13	0.15	0.21	0.12	0.17	0.16	0.17	0.00

The United States has the lowest risk, around 17%. The next lowest risk country is Canada. This is partly due to the fact that in U.S. dollar terms, U.S. stocks have no currency exposure, and the Canadian dollar is closely linked to the U.S. dollar. The two markets that have the highest risks are Hong Kong and Singapore, with risks above 44%. The economy and stock market in both regions are rather concentrated in the financial sector. All other markets have risks between 25% and 35%. The Sharpe ratio is all positive except Italy, which has a Sharpe ratio near zero. The United States has the best Sharpe ratio at 0.29 because of its low risk. The other eight markets have Sharpe ratios between 0.12 and 0.21. We note that these Sharpe ratios are achieved over the span of 47 years.

Table 9.2 shows the pairwise differences among the countries. The differences are generally moderate except for pairs involving Hong Kong and Italy. This is because Hong Kong has the highest return and Italy has the lowest return, while all other countries' returns are similar.

Table 9.3 shows the volatility of return differences among all pairs. The volatility is the highest between Hong Kong, Singapore, and non-Asian countries. Hong Kong and Singapore have the highest return volatilities, and they also have relatively lower correlations with non-Asian countries. Since they have relatively higher correlations with Japan and Australia, the volatility of return differences between these four countries is not high.

Table 9.4 shows the pairwise t-statistics of all pairs. Recall that if the absolute value of the t-statistic is less than 1, the rebalancing alpha of a long-only portfolio involving the pair is positive. On the other hand, if the absolute value of the t-statistic is greater than 1, the rebalancing alpha of a long-only portfolio involving the pair is negative. Out of 45 unique pairs, there are only 5 pairs with negative rebalancing alpha, 4 of which involve Italy. This is because its return is so much lower than that of other countries. The remaining pair is between Hong Kong and Australia. We also note that since the returns of both Hong Kong and Italy are rather different from those of other countries, absolute values of t-statistics of pairs involving the two countries are generally higher than those of other pairs.

We choose an equally weighted portfolio of the 10 countries to study the rebalancing alpha. It should be noted that since most of the t-statistics in Table 9.4 are between −1 and 1, any long-only portfolio would most likely have positive rebalancing alpha. We calculate returns of the fixed-weight (FW) and buy-and-hold (BH) portfolios using actual annual returns. We also derive approximated volatility and return effects using analytic results derived in previous chapters. Both actual results

TABLE 9.2 Pairwise Return Difference among Asset Classes

	United States	Canada	Japan	Australia	Hong Kong	Singapore	United Kingdom	Germany	France	Italy
United States	0.00%	0.74%	0.92%	1.19%	-4.26%	-0.36%	0.47%	0.45%	0.34%	5.07%
Canada	-0.74%	0.00%	0.18%	0.44%	-5.00%	-1.10%	-0.28%	-0.29%	-0.41%	4.33%
Japan	-0.92%	-0.18%	0.00%	0.26%	-5.18%	-1.28%	-0.46%	-0.47%	-0.59%	4.14%
Australia	-1.19%	-0.44%	-0.26%	0.00%	-5.44%	-1.55%	-0.72%	-0.74%	-0.85%	3.88%
Hong Kong	4.26%	5.00%	5.18%	5.44%	0.00%	3.90%	4.72%	4.71%	4.59%	9.33%
Singapore	0.36%	1.10%	1.28%	1.55%	-3.90%	0.00%	0.83%	0.81%	0.70%	5.43%
United Kingdom	-0.47%	0.28%	0.46%	0.72%	-4.72%	-0.83%	0.00%	-0.02%	-0.13%	4.60%
Germany	-0.45%	0.29%	0.47%	0.74%	-4.71%	-0.81%	0.02%	0.00%	-0.12%	4.62%
France	-0.34%	0.41%	0.59%	0.85%	-4.59%	-0.70%	0.13%	0.12%	0.00%	4.73%
Italy	-5.07%	-4.33%	-4.14%	-3.88%	-9.33%	-5.43%	-4.60%	-4.62%	-4.73%	0.00%

TABLE 9.3 Pairwise Volatility of Return Differences among Asset Classes

	United States	Canada	Japan	Australia	Hong Kong	Singapore	United Kingdom	Germany	France	Italy
United States	0.00%	17.17%	30.54%	21.42%	38.30%	41.13%	20.46%	23.65%	21.43%	29.93%
Canada	17.17%	0.00%	30.28%	16.96%	36.01%	35.90%	24.85%	27.44%	23.28%	33.19%
Japan	30.54%	30.28%	0.00%	31.54%	36.45%	30.20%	34.09%	33.22%	29.29%	34.98%
Australia	21.42%	16.96%	31.54%	0.00%	35.16%	36.49%	22.15%	28.69%	21.70%	30.08%
Hong Kong	38.30%	36.01%	36.45%	35.16%	0.00%	28.49%	37.17%	44.10%	39.22%	45.36%
Singapore	41.13%	35.90%	30.20%	36.49%	28.49%	0.00%	42.33%	46.86%	42.30%	50.15%
United Kingdom	20.46%	24.85%	34.09%	22.15%	37.17%	42.33%	0.00%	27.40%	24.76%	33.75%
Germany	23.65%	27.44%	33.22%	28.69%	44.10%	46.86%	27.40%	0.00%	18.78%	25.93%
France	21.43%	23.28%	29.29%	21.70%	39.22%	42.30%	24.76%	18.78%	0.00%	21.63%
Italy	29.93%	33.19%	34.98%	30.08%	45.36%	50.15%	33.75%	25.93%	21.63%	0.00%

TABLE 9.4 Pairwise t-Statistics among Different Pairs

	United States	Canada	Japan	Australia	Hong Kong	Singapore	United Kingdom	Germany	France	Italy
United States		0.29	0.21	0.38	-0.75	-0.06	0.15	0.13	0.11	**1.15**
Canada	-0.29		0.04	0.18	-0.94	-0.21	-0.08	-0.07	-0.12	0.88
Japan	-0.21	-0.04		0.06	-0.96	-0.29	-0.09	-0.10	-0.14	0.80
Australia	-0.38	-0.18	-0.06		**-1.05**	-0.29	-0.22	-0.17	-0.27	0.88
Hong Kong	0.75	0.94	0.96	**1.05**		0.93	0.86	0.72	0.79	**1.39**
Singapore	0.06	0.21	0.29	0.29	-0.93		0.13	0.12	0.11	0.73
United Kingdom	-0.15	0.08	0.09	0.22	-0.86	-0.13		0.00	-0.04	0.92
Germany	-0.13	0.07	0.10	0.17	-0.72	-0.12	0.00		-0.04	**1.21**
France	-0.11	0.12	0.14	0.27	-0.79	-0.11	0.04	0.04		**1.48**
Italy	**-1.15**	-0.88	-0.80	-0.88	**-1.39**	-0.73	-0.92	**-1.21**	**-1.48**	

Those whose absolute value is above 1 are in bold face.

TABLE 9.5 Returns of FW and BH Portfolios Based on the Equally Weighted Portfolio and Volatility, Return Effects, and Rebalancing Alpha

	g_{FW}	g_{BH}	$\bar{g}$	e_v	e_r	α
Actual	11.11%	10.58%	9.56%	1.55%	1.02%	0.53%
Approximation				2.34%	1.04%	1.30%

and approximations of the equally weighted portfolio are displayed in Table 9.5. The return of the FW portfolio is higher than that of the BH portfolio by 53 basis points (bps), resulting in a positive rebalancing alpha. Both volatility and return effects are positive at 1.55% and 1.02%, respectively. The approximated volatility and return effects are 2.34% and 1.04%, respectively. This indicates a rebalancing alpha of 130 bps. The error for the volatility effect is quite large, while the error of the return effect is just 2 bps. The reason for the error in the estimation of volatility effect is likely to be the high volatilities of Hong Kong and Singapore. Indeed, when we exclude the two countries from the portfolio, the error declines significantly.

Figure 9.1 displays the average serial cross-correlations of the 10 countries as well as their scaled contribution over the period. From 1 to 12 years, the correlations are on average negative, indicating mean reversion. From 12 to 24 years, the correlations are mostly positive, indicating cross-sectional momentum. The rest of the correlations, from 24 years on, are quite random. Overall, the scaled contributions from serial cross-sectional correlations are negative. This would lead to a positive rebalancing alpha for this period.

This fact is reflected in Figure 9.2, which shows the cumulative contribution from cross-sectional serial correlations to the variance of arithmetic means. Recall that the return effect is proportional to the cross-sectional variance of geometric means, which is highly dependent on the cross-sectional variance of arithmetic means. Figure 9.2 shows that the cross-sectional serial correlations contribute negatively to the variance of arithmetic means over this period. It is not surprising that the rebalancing alpha is positive.

Portfolio rebalancing alpha depends on portfolio weights as well as on investment horizon. We have shown that for the equally weighted portfolio, the rebalancing alpha is +53 bps for the entire period from 1970 to 2016. Next, we compute returns of FW and BH portfolios with a 15-year investment horizon, and the results are shown in Figure 9.3. The starting year of the investment period is indicated on the x-axis. For instance,

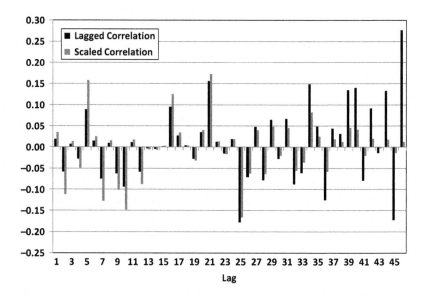

FIGURE 9.1   Average cross-sectional serial correlation of asset returns for different lags and their scaled contribution to the variance of arithmetic means.

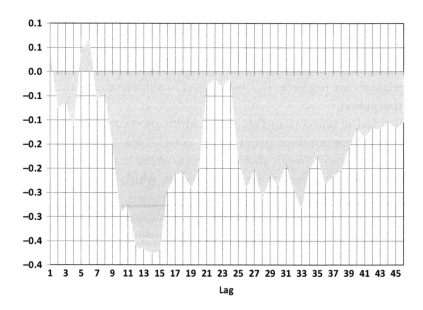

FIGURE 9.2   Cumulative contributions of cross-sectional serial correlations to the variance reduction of arithmetic means.

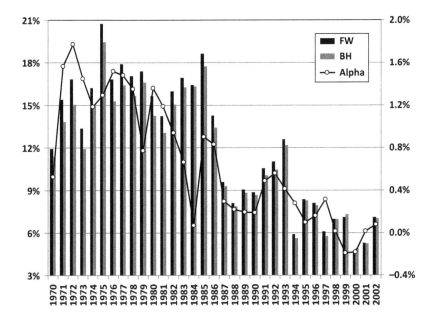

FIGURE 9.3   Returns of FW and BH portfolios and rebalancing alphas for an investment horizon of 15 years.

the year 1970 indicates the period from 1970 to 1984, and the year 2002 indicates the period from 2002 to 2016. The rebalancing alpha is generally positive, but it has been declining steadily over time. It has become slightly negative in the periods starting 1999 and 2000. In the periods starting in the 1980s, the rebalancing alpha is as high as 100 bps. Of course, this decrease in portfolio rebalancing reflects changes in return characteristics in equity market returns.

- First, a decrease in equity return volatilities leads to a decrease in volatility effects.

- On further inspection, we note that cross-sectional correlations have become positive in the most recent period, leading to a relative increase in return effects.

- From a pairwise perspective, there are more pairs whose t-statistics are either lower than −1 or greater than +1, and they contribute to negative rebalancing alpha.

Lastly, we also notice from Figure 9.3 that the returns of the equally weighted portfolio have declined in recent periods. For example, returns in the 1970s and early 1980s were above 12% for both FW and BH portfolios, while returns since 1994 were all below 9%.

9.2 BOND PORTFOLIOS

The country returns we use for bond portfolios are Citigroup World Government Bond Index (WGBI) country index returns in U.S. dollar hedged terms. Unlike equity returns, the bond returns left unhedged will be dominated by currency returns. Therefore, it is important to use hedged returns to analyze return and risk of government bond returns from different countries. There are over 20 developed countries in the WGBI world index. For the purpose of illustration, we shall select 10 countries for our analysis. They represent major economies of the world and also major countries in terms of index weights. Their U.S. dollar hedged return series go back to 1986. Other countries in the index do not go back that far.

Table 9.6 displays annualized return, standard deviation, and Sharpe ratio of the 10 countries, based on annual returns from 1986 to 2016. They are the United States, Canada, Japan, Australia, the United Kingdom, Germany, France, Italy, the Netherlands, and Switzerland. The annualized geometric returns are mostly close together in a range of only 100 bps, all between 5.76% and 6.78%. The risk or standard deviation of returns does not differ much more across countries, with Switzerland having the lowest risk at 4.54% and Australia having the highest risk at 6.87%. The Sharpe ratios are all positive, between 0.35 for Australia and 0.67 for Japan. We make two comments about these Sharpe ratios,

- The period from 1986 to 2016 saw the declines of interest rates across the world. As a result, government bonds have achieved higher Sharpe ratios than their historical averages as well as stocks' Sharpe ratios, as shown in Section 9.1.

- The expectation is that as global interest rates rise in the future, the Sharpe ratios of government bonds will mean-revert toward their long-term averages between 0.3 and 0.4. However, the timing of this mean-reverting process is uncertain. Many investors have been calling for rising interest rates for years, and those predictions have been consistently too early or wrong.

TABLE 9.6   Return Statistics of 10 Equity Markets Based on Annual Returns from 1970 to 2016

	United States	Canada	Japan	Australia	United Kingdom	Germany	France	Italy	Netherlands	Switzerland
Return	6.29%	6.50%	6.52%	5.76%	6.35%	6.22%	6.58%	6.78%	6.45%	5.80%
Risk	6.04%	5.80%	4.73%	6.87%	6.64%	4.58%	5.43%	6.47%	5.00%	4.54%
Sharpe	0.48	0.54	0.67	0.35	0.45	0.62	0.59	0.53	0.62	0.53

Table 9.7 shows the pairwise differences among the countries. The differences are generally small, with those involving Australia, Switzerland, and Italy being slightly higher.

Table 9.8 shows the volatility of return differences among all pairs. The volatility is generally low except for pairs involving Italy. The reason for this exception is that compared with other countries, Italian government bonds have much higher credit risk than other sovereign rates. This is especially true after the global financial crisis of 2008. The credit risk exposure causes Italian government bonds to have significant positive exposures to equity markets and lower correlations with government bonds of other countries, resulting in higher pairwise volatilities as shown in Table 9.8.

Table 9.9 shows the pairwise t-statistics of all pairs. Out of 45 unique pairs, there are only 4 pairs with negative rebalancing alpha, 3 of which involve Switzerland, which has the second lowest return and the lowest risk. The remaining pair is between Canada and Australia.

Similarly, we choose an equally weighted portfolio of the 10 countries to study the rebalancing alpha. It should be noted that since most of the t-statistics in Table 9.9 are between −1 and 1, any long-only portfolio would most likely have positive rebalancing alpha. We calculate returns of the FW and BH portfolios using actual annual returns. We also derive approximated volatility and return effects using analytic results derived in previous chapters. Both actual results and approximations of the equally weighted portfolio are displayed in Table 9.10. The return of the FW portfolio is higher than that of the BH portfolio by just 3.1 bps, resulting in a positive rebalancing alpha. Both volatility and return effects are marginally positive at 4.4 and 1.3 bps, respectively. The approximated volatility and return effects are 4.9 and 1.4 bps, respectively. Compared with the equity case, the magnitudes of volatility and return effects as well as rebalancing alpha are much smaller. This is because for the government bonds, the hedged returns have low volatilities, resulting in a low volatility effect. In addition, the returns are not too dissimilar, resulting in a low return effect. In other words, for the BH portfolio, there will be little drift over time; hence, little rebalancing is required for the FW portfolio. It can be concluded that rebalancing alpha, whether positive or negative, will be insignificant.

Figure 9.4 displays the average serial cross-correlations of the 10 countries as well as their scaled contribution over the period. From one to four years, the correlations are on average negative, indicating mean reversion. From five to eight years, the correlations are mostly positive, indicating

TABLE 9.7 Pairwise Return Difference among Asset Classes

	United States	Canada	Japan	Australia	United Kingdom	Germany	France	Italy	Netherlands	Switzerland
United States	0.00%	−0.21%	−0.23%	0.52%	−0.06%	0.07%	−0.29%	−0.49%	−0.16%	0.49%
Canada	0.21%	0.00%	−0.01%	0.74%	0.15%	0.28%	−0.08%	−0.28%	0.05%	0.71%
Japan	0.23%	0.01%	0.00%	0.75%	0.17%	0.30%	−0.07%	−0.26%	0.07%	0.72%
Australia	−0.52%	−0.74%	−0.75%	0.00%	−0.59%	−0.46%	−0.82%	−1.02%	−0.69%	−0.03%
United Kingdom	0.06%	−0.15%	−0.17%	0.59%	0.00%	0.13%	−0.23%	−0.43%	−0.10%	0.55%
Germany	−0.07%	−0.28%	−0.30%	0.46%	−0.13%	0.00%	−0.36%	−0.56%	−0.23%	0.42%
France	0.29%	0.08%	0.07%	0.82%	0.23%	0.36%	0.00%	−0.20%	0.13%	0.79%
Italy	0.49%	0.28%	0.26%	1.02%	0.43%	0.56%	0.20%	0.00%	0.33%	0.99%
Netherlands	0.16%	−0.05%	−0.07%	0.69%	0.10%	0.23%	−0.13%	−0.33%	0.00%	0.65%
Switzerland	−0.49%	−0.71%	−0.72%	0.03%	−0.55%	−0.42%	−0.79%	−0.99%	−0.65%	0.00%

TABLE 9.8 Pairwise Volatility of Return Differences among Asset Classes

	United States	Canada	Japan	Australia	United Kingdom	Germany	France	Italy	Netherlands	Switzerland
United States	0.00%	3.14%	**5.13%**	4.06%	5.50%	4.13%	4.88%	**7.78%**	4.67%	4.90%
Canada	3.14%	0.00%	**4.18%**	3.54%	4.85%	3.31%	3.71%	**6.56%**	3.71%	3.90%
Japan	**5.13%**	**4.18%**	**0.00%**	**5.03%**	**5.55%**	**3.78%**	**4.56%**	**6.40%**	**3.89%**	**3.76%**
Australia	4.06%	3.54%	**5.03%**	0.00%	4.28%	4.12%	4.65%	**7.15%**	4.29%	4.95%
United Kingdom	5.50%	4.85%	**5.55%**	4.28%	0.00%	3.55%	4.82%	**7.56%**	3.48%	4.48%
Germany	4.13%	3.31%	**3.78%**	4.12%	3.55%	0.00%	2.69%	**6.00%**	1.17%	2.34%
France	4.88%	3.71%	**4.56%**	4.65%	4.82%	2.69%	0.00%	**4.54%**	2.60%	3.67%
Italy	**7.78%**	**6.56%**	**6.40%**	**7.15%**	**7.56%**	**6.00%**	**4.54%**	**0.00%**	**5.76%**	**6.15%**
Netherlands	4.67%	3.71%	**3.89%**	4.29%	3.48%	1.17%	2.60%	**5.76%**	0.00%	2.47%
Switzerland	4.90%	3.90%	**3.76%**	4.95%	4.48%	2.34%	3.67%	**6.15%**	2.47%	0.00%

Those whose absolute value is above 1 are in bold face.

TABLE 9.9 Pairwise t-Statistics among Different Pairs

	United States	Canada	Japan	Australia	United Kingdom	Germany	France	Italy	Netherlands	Switzerland
United States	0.00	-0.37	-0.24	0.71	-0.06	0.09	-0.33	-0.35	-0.19	0.55
Canada	0.37	0.00	-0.02	1.14	0.17	0.47	-0.12	-0.23	0.08	0.99
Japan	0.24	0.02	0.00	0.82	0.17	0.43	-0.08	-0.23	0.10	1.05
Australia	-0.71	-1.14	-0.82	0.00	-0.75	-0.61	-0.96	-0.78	-0.88	-0.04
United Kingdom	0.06	-0.17	-0.17	0.75	0.00	0.20	-0.26	-0.31	-0.16	0.68
Germany	-0.09	-0.47	-0.43	0.61	-0.20	0.00	-0.74	-0.51	-1.07	0.99
France	0.33	0.12	0.08	0.96	0.26	0.74	0.00	-0.24	0.28	1.17
Italy	0.35	0.23	0.23	0.78	0.31	0.51	0.24	0.00	0.32	0.88
Netherlands	0.19	-0.08	-0.10	0.88	0.16	1.07	-0.28	-0.32	0.00	1.45
Switzerland	-0.55	-0.99	-1.05	0.04	-0.68	-0.99	-1.17	-0.88	-1.45	0.00

Those whose absolute value is above 1 are in bold face.

TABLE 9.10   Returns of FW and BH Portfolios Based on the Equally Weighted Portfolio and Volatility, Return Effects, and Rebalancing Alpha

	g_{FW}	g_{BH}	$\bar{g}$	e_v	e_r	α
Actual	6.37%	6.34%	6.33%	0.044%	0.013%	0.031%
Approximation				0.049%	0.014%	0.035%

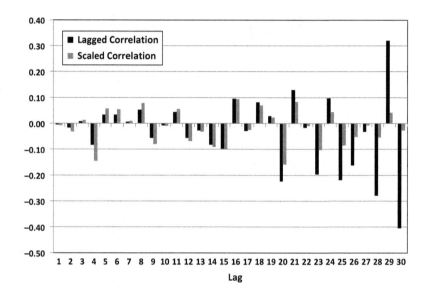

FIGURE 9.4   Average cross-sectional serial correlation of asset returns for different lags and their scaled contribution to the variance of arithmetic means.

cross-sectional momentum. The rest of the correlations, from nine years on, are again negative in general. Overall, the scaled contributions from serial cross-sectional correlations are negative. This would lead to a positive rebalancing alpha for this period.

This fact is reflected in Figure 9.5, which shows the cumulative contribution from cross-sectional serial correlations to the variance of arithmetic means. Recall that the return effect is proportional to the cross-sectional variance of geometric means, which is highly dependent on the cross-sectional variance of arithmetic means. Figure 9.5 shows that cross-sectional serial correlations contribute negatively to the variance of arithmetic means over this period. The rebalancing alpha is positive, albeit small.

Next, we compute returns of FW and BH portfolios with a 15-year investment horizon, and the results are shown in Figure 9.6. The starting year of the investment period is indicated on the x-axis. For instance, the

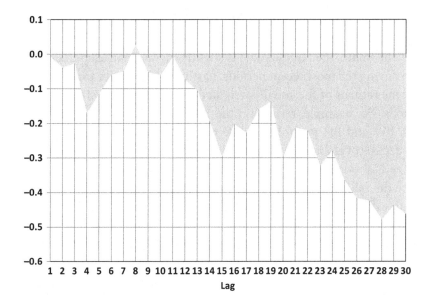

FIGURE 9.5   Cumulative contributions of cross-sectional serial correlations to the variance reduction of arithmetic means.

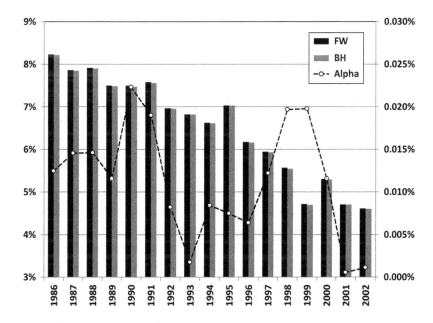

FIGURE 9.6   Returns of FW and BH portfolios and rebalancing alphas for an investment horizon of 15 years.

year 1986 indicates the period from 1986 to 2000, and the year 2002 indicates the period from 2002 to 2016. The rebalancing alpha is generally positive but small. The largest value is slightly higher than 2 bps. It is nearly zero during the most recent periods. Lastly, we also notice from Figure 9.6 that the returns of the equally weighted portfolio have declined in recent periods. For example, returns in the 1980s were between 7% and 8% for both FW and BH portfolios, while returns since 1999 were around 5%. The decline of global interest rates is behind the trend of declining returns.

The FW portfolios have outperformed the BH portfolios consistently. Since the rebalancing alpha is not economically significant, it is interesting to ask whether the outperformance is statistically significant, which would be the case if the tracking error between the FW and the BH portfolios were very low. Further analysis shows that this is not the case. We found that the average tracking error over all the 15-year periods shown in Figure 9.6 is around 6 bps, and the average information ratio (IR), defined as the ratio of alpha to the tracking error, is less than 0.2.

9.3 COMMODITY PORTFOLIOS

The commodity returns we use for commodity portfolios are Standard and Poor's Goldman Sachs Commodity Index (GSCI) total returns in U.S. dollar terms. There are over 20 commodities in the index. For the purpose of illustration, we shall select 10 commodities for our analysis. They represent commodity sectors: energy, precious metal, industrial metal, live stocks, grains, and softs. Their return series go back to 1989.

Table 9.11 displays the annualized return, standard deviation, and Sharpe ratio of the 10 commodities, based on annual returns from 1989 to 2016. They are Crude Oil, Heating Oil, Gold, Copper, Live Cattle, Lean Hogs, Corn, Soybeans, Cotton, and Sugar. The annualized geometric returns are very diverse, ranging from close to −6% to 8%. The risks or standard deviations of returns show the same pattern, ranging from 13% to 47%. The Sharpe ratio is generally low, even when it is positive. We make two comments on commodity returns.

- While stocks and bonds possess risk premiums based on sound economic argument, it is debatable whether commodities as an asset class have risk premium over cash. If the risk premium were zero, then the long-term Sharpe ratio of commodity returns would be zero. Empirically, this seems to be the case. While stocks and bonds have a long-term Sharpe ratio near 0.3, commodities in general have a much lower Sharpe ratio.

TABLE 9.11 Return Statistics of 10 Equity Markets Based on Annual Returns from 1989 to 2016

	Crude Oil	Heating Oil	Gold	Copper	Live Cattle	Lean Hogs	Corn	Soybeans	Cotton	Sugar
Return	4.69%	5.20%	3.72%	7.91%	2.19%	−5.79%	−5.85%	4.56%	−1.03%	3.65%
Risk	46.91%	41.10%	14.22%	39.30%	12.83%	22.49%	21.84%	23.18%	30.40%	30.87%
Sharpe	0.03	0.05	0.05	0.12	−0.07	−0.39	−0.41	0.06	−0.13	0.02

- Commodities do provide a hedge against rising inflation. This was apparent in the 1970s. However, a detailed study of the commodity market during that period is beyond the scope of this book.

Table 9.12 shows the pairwise differences among the 10 commodities.

Table 9.13 shows the volatility of return differences among all pairs. The volatility is generally high except for pairs involving Gold and Live Cattle. Also, commodities within the same sector have lower volatility of return differences. Examples are Crude Oil/Heating Oil and Corn/Soybeans.

Table 9.14 shows the pairwise *t*-statistics of all pairs. Out of 45 unique pairs, there are 16 pairs with negative rebalancing alpha, 10 of which involve Lean Hogs and Corn. Both Lean Hogs and Corn have very low returns compared with other commodities.

Similarly, we choose an equally weighted portfolio of the 10 countries to study the rebalancing alpha. We calculate returns of the FW and BH portfolios using actual annual returns. We also derive approximated volatility and return effects using analytic results derived in previous chapters. Both actual results and approximations of the equally weighted portfolio are displayed in Table 9.15. The return of the FW portfolio is higher than that of the BH portfolio by 68 bps, resulting in a positive rebalancing alpha. Both volatility and return effects are quite large, at 251 and 184 bps, respectively. The approximated volatility and return effects are 315 and 265 bps, respectively, and both overestimate the true values. Compared with stocks and bonds, the magnitudes of volatility and return effects are much larger due to high return volatilities and differentiating returns. However, the combined rebalancing alpha is quite modest.

Figure 9.7 displays the average serial cross-correlations of the 10 commodities as well as their scaled contribution over the period. From one to three years, the correlations are on average negative, indicating mean reversion. From four to six years, the correlations are mostly positive, indicating cross-sectional momentum. The correlations from 17 years on are again negative in general. Overall, the scaled contributions from serial cross-sectional correlations are negative. This would lead to a positive rebalancing alpha for this period.

This fact is reflected in Figure 9.8, which shows the cumulative contribution from cross-sectional serial correlations to the variance of arithmetic means. Recall that the return effect is proportional to the cross-sectional

TABLE 9.12 Pairwise Return Difference among Asset Classes

	Crude Oil	Heating Oil	Gold	Copper	Live Cattle	Lean Hogs	Corn	Soybeans	Cotton	Sugar
Crude Oil	0.00%	−0.50%	0.97%	−3.22%	2.50%	10.48%	10.54%	0.14%	5.73%	1.04%
Heating Oil	0.50%	0.00%	1.48%	−2.71%	3.01%	10.98%	11.05%	0.64%	6.23%	1.55%
Gold	−0.97%	−1.48%	0.00%	−4.19%	1.53%	9.51%	9.57%	−0.83%	4.76%	0.07%
Copper	3.22%	2.71%	4.19%	0.00%	5.72%	13.70%	13.76%	3.36%	8.95%	4.26%
Live Cattle	−2.50%	−3.01%	−1.53%	−5.72%	0.00%	7.98%	8.04%	−2.37%	3.22%	−1.46%
Lean Hogs	−10.48%	−10.98%	−9.51%	−13.70%	−7.98%	0.00%	0.06%	−10.34%	−4.75%	−9.44%
Corn	−10.54%	−11.05%	−9.57%	−13.76%	−8.04%	−0.06%	0.00%	−10.40%	−4.81%	−9.50%
Soybeans	−0.14%	−0.64%	0.83%	−3.36%	2.37%	10.34%	10.40%	0.00%	5.59%	0.90%
Cotton	−5.73%	−6.23%	−4.76%	−8.95%	−3.22%	4.75%	4.81%	−5.59%	0.00%	−4.69%
Sugar	−1.04%	−1.55%	−0.07%	−4.26%	1.46%	9.44%	9.50%	−0.90%	4.69%	0.00%

TABLE 9.13   Pairwise Volatility of Return Differences among Asset Classes

	Crude Oil	Heating Oil	Gold	Copper	Live Cattle	Lean Hogs	Corn	Soybeans	Cotton	Sugar
Crude Oil	0.00%	15.35%	47.84%	51.34%	42.54%	41.07%	52.59%	51.13%	49.76%	47.28%
Heating Oil	15.35%	0.00%	41.07%	46.16%	37.24%	37.35%	47.47%	44.24%	44.07%	38.20%
Gold	47.84%	41.07%	0.00%	36.09%	18.82%	29.08%	20.86%	20.04%	29.80%	31.82%
Copper	51.34%	46.16%	36.09%	0.00%	38.45%	45.04%	44.48%	41.40%	39.46%	36.32%
Live Cattle	42.54%	37.24%	18.82%	38.45%	0.00%	17.67%	23.95%	25.34%	29.52%	32.11%
Lean Hogs	41.07%	37.35%	29.08%	45.04%	17.67%	0.00%	30.37%	36.11%	39.35%	36.59%
Corn	52.59%	47.47%	20.86%	44.48%	23.95%	30.37%	0.00%	22.95%	29.40%	38.09%
Soybeans	51.13%	44.24%	20.04%	41.40%	25.34%	36.11%	22.95%	0.00%	28.99%	35.86%
Cotton	49.76%	44.07%	29.80%	39.46%	29.52%	39.35%	29.40%	28.99%	0.00%	34.24%
Sugar	47.28%	38.20%	31.82%	36.32%	32.11%	36.59%	38.09%	35.86%	34.24%	0.00%

TABLE 9.14 Pairwise t-Statistics among Different Pairs

	Crude Oil	Heating Oil	Gold	Copper	Live Cattle	Lean Hogs	Corn	Soybeans	Cotton	Sugar
Crude Oil	0.00	−0.17	0.11	−0.33	0.31	1.33	1.04	0.01	0.60	0.11
Heating Oil	0.17	0.00	0.19	−0.31	0.42	1.53	1.21	0.08	0.73	0.21
Gold	−0.11	−0.19	0.00	−0.60	0.42	1.70	2.38	−0.22	0.83	0.01
Copper	0.33	0.31	0.60	0.00	0.77	1.58	1.61	0.42	1.18	0.61
Live Cattle	−0.31	−0.42	−0.42	−0.77	0.00	2.35	1.74	−0.49	0.57	−0.24
Lean Hogs	−1.33	−1.53	−1.70	−1.58	−2.35	0.00	0.01	−1.49	−0.63	−1.34
Corn	−1.04	−1.21	−2.38	−1.61	−1.74	−0.01	0.00	−2.35	−0.85	−1.30
Soybeans	−0.01	−0.08	0.22	−0.42	0.49	1.49	2.35	0.00	1.00	0.13
Cotton	−0.60	−0.73	−0.83	−1.18	−0.57	0.63	0.85	−1.00	0.00	−0.71
Sugar	−0.11	−0.21	−0.01	−0.61	0.24	1.34	1.30	−0.13	0.71	0.00

TABLE 9.15   Returns of FW and BH Portfolios Based on the Equally Weighted Portfolio and Volatility, Return Effects, and Rebalancing Alpha

	g_{FW}	g_{BH}	$\bar{g}$	e_v	e_r	α
Actual	4.44%	3.76%	1.92%	2.51%	1.84%	0.68%
Approximation				3.15%	2.65%	0.50%

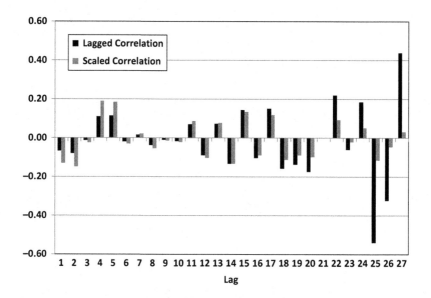

FIGURE 9.7   Average cross-sectional serial correlation of asset returns for different lags and their scaled contribution to the variance of arithmetic means.

variance of geometric means, which is highly dependent on the cross-sectional variance of arithmetic means. Figure 9.8 shows that cross-sectional serial correlations contribute negatively to the variance of arithmetic means over this period. The rebalancing alpha is positive.

Next, we compute returns of FW and BH portfolios with a 15-year investment horizon, and the results are shown in Figure 9.9. The starting year of the investment period is indicated on the x-axis. For instance, the year 1989 indicates the period from 1989 to 2003, and the year 2002 indicates the period from 2002 to 2016. The rebalancing alpha is positive initially but then turns negative. It is close to −100 bps in most recent periods. Lastly, we also notice from Figure 9.9 that the returns of the equally weighted portfolio have declined in recent periods.

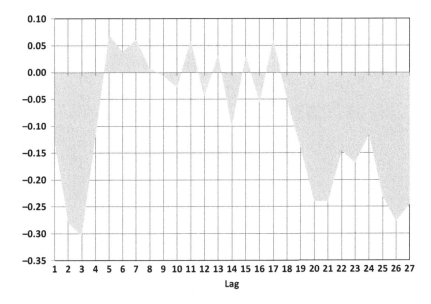

FIGURE 9.8   Cumulative contributions of cross-sectional serial correlations to the variance reduction of arithmetic means.

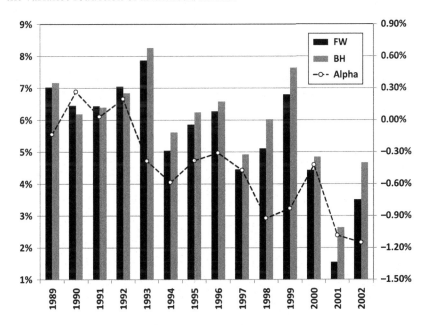

FIGURE 9.9   Returns of FW and BH portfolios and rebalancing alphas for an investment horizon of 15 years.

CHAPTER 10

Rebalancing Alpha and Mean Reversion

Previously, we have analyzed both the volatility and the return effects of portfolio rebalancing and then combined the results together to derive an approximation of rebalancing alpha. In this chapter, we take a direct approach by deriving the terminal wealth of the fixed-weight (FW) and buy-and-hold (BH) portfolios without approximation. As in the previous analysis, we first gain insights by examining the case of two-asset portfolios.

10.1 TWO-ASSET TWO-PERIOD CASE

Consider the terminal wealth of an FW and a BH portfolio with two assets and after two periods. The FW portfolio's terminal wealth is

$$W_{FW} = \left(1 + w_1 r_{11} + w_2 r_{21}\right)\left(1 + w_1 r_{12} + w_2 r_{22}\right)$$
$$= \left[w_1\left(1+r_{11}\right) + w_2\left(1+r_{21}\right)\right]\left[w_1\left(1+r_{12}\right) + w_2\left(1+r_{22}\right)\right] \tag{10.1}$$

The terminal wealth of the BH portfolio is

$$W_{BH} = w_1\left(1+r_{11}\right)\left(1+r_{12}\right) + w_2\left(1+r_{21}\right)\left(1+r_{22}\right) \tag{10.2}$$

The difference in the terminal wealth is

$$W_{FW} - W_{BH} = \left[w_1 \left(1+r_{11}\right) + w_2 \left(1+r_{21}\right) \right] \left[w_1 \left(1+r_{12}\right) + w_2 \left(1+r_{22}\right) \right]$$
$$-w_1 \left(1+r_{11}\right)\left(1+r_{12}\right) - w_2 \left(1+r_{21}\right)\left(1+r_{22}\right) \tag{10.3}$$

Equation 10.3 can be simplified to yield

$$W_{FW} - W_{BH} = -w_1 w_2 \left(r_{11} - r_{21}\right)\left(r_{12} - r_{22}\right) \tag{10.4}$$

The return differences on the right-hand side of the equation are the return spreads of the two assets in Periods 1 and 2.

Equation 10.4 is an exact expression with no approximations. For a long-only portfolio in which both weights are positive, the FW portfolio outperforms the BH portfolio if

$$\left(r_{11} - r_{21}\right)\left(r_{12} - r_{22}\right) < 0 \tag{10.5}$$

Therefore, the return differences for the two periods must be of opposite signs. For example, if Asset 1 outperforms in Period 1, then Asset 2 will have to outperform in Period 2, and vice versa. This is cross-sectional return reversal.

On the other hand, if the portfolio is a long-short portfolio, then $w_1 w_2 < 0$. In this case, for the FW portfolio to outperform the BH portfolio, we must have

$$\left(r_{11} - r_{21}\right)\left(r_{12} - r_{22}\right) > 0 \tag{10.6}$$

The return differences are of the same sign. The winning asset in Period 1 continues to be the winning asset in Period 2, and the losing asset in Period 1 continues to be the losing asset in Period 2. This is a form of momentum.

Example 10.1: We consider a portfolio of two assets with 50% invested in each. The period returns are displayed in Table 10.1. In Period 1, returns are 5% for Asset 1 and 10% for Asset 2, while in Period 2, returns are 0% for Asset 1 and −10% for Asset 2. For the FW portfolio, the returns are 7.5% for Period 1 and −5% for Period 2. So, its terminal wealth is $W_{FW} = \left(1+7.5\%\right)\left(1-5\%\right) = 1.02125$. On the

TABLE 10.1   Portfolio Weights and Period Returns of Two Assets

	Weight	Period 1	Period 2	Wealth
Asset 1	50%	5%	0%	1.05
Asset 2	50%	10%	−10%	0.99
FW return		7.5%	−5%	

other hand, the terminal wealth of two assets invested on their own is 1.05 and 0.99, respectively. Hence, the terminal wealth of the BH portfolio is the average of the two due to the 50/50 weights. We have $W_{BH} = 0.5 \cdot 1.05 + 0.5 \cdot 0.99 = 1.02$. Therefore, the FW portfolio outperforms the BH portfolio; $W_{FW} - W_{BH} = 0.00125$.

We note that the asset returns are mean-reverting in this example, since the return difference between Assets 1 and 2 is −5% in Period 1, while the difference is 10% in Period 2. The two are of different sign. We can calculate the rebalancing alpha by Equation 10.4, i.e. $W_{FW} - W_{BH} = -0.5 \cdot 0.5 \cdot (-0.05) \cdot 0.1 = 0.00125$. We have the same number derived previously with the terminal wealth of the FW and BH portfolios.

Suppose we assume that the return difference between the two assets follows a normal distribution $N(\mu, \sigma^2)$, and the serial correlation between two subsequent periods is $\rho_{1,2}$. Then, the expectation of Equation 10.4 yields

$$E(W_{FW} - W_{BH}) = -w_1 w_2 (\mu^2 + \rho_{1,2}\sigma^2) \tag{10.7}$$

Hence, for a long-only portfolio, rebalancing is expected to add value if

$$\mu^2 + \rho_{1,2}\sigma^2 < 0 \text{ or } \rho_{1,2} < -\left(\frac{\mu}{\sigma}\right)^2 \tag{10.8}$$

In other words, the serial autocorrelation of the return difference has to be less than the risk-adjusted return squared. It definitely needs to be negative. Otherwise, portfolio rebalancing detracts value. For a long-short portfolio, the condition is just the opposite.

10.2 MULTIPLE-ASSET TWO-PERIOD CASE

The previous case can be generalized to M assets for two investment periods. Suppose the portfolio weights are w_i, $i = 1, \cdots, M$, and asset returns in the first period are

$$\mathbf{r}_1 = (r_{11}, r_{21}, \cdots, r_{M1})' \tag{10.9}$$

Asset returns in the second period are

$$\mathbf{r}_2 = (r_{12}, r_{22}, \cdots, r_{M2})' \tag{10.10}$$

Both sets of returns are written as vectors.

The terminal wealth of the FW portfolio at the end of two periods is

$$
\begin{aligned}
W_{\mathrm{FW}} &= \left(1 + \sum_{i=1}^{M} w_i r_{i1}\right)\left(1 + \sum_{i=1}^{M} w_i r_{i2}\right) \\
&= 1 + \sum_{i=1}^{M} w_i r_{i1} + \sum_{i=1}^{M} w_i r_{i2} + \left(\sum_{i=1}^{M} w_i r_{i1}\right)\left(\sum_{i=1}^{M} w_i r_{i2}\right)
\end{aligned} \tag{10.11}
$$

The terminal wealth of the BH portfolio is

$$
\begin{aligned}
W_{\mathrm{BH}} &= \sum_{i=1}^{M} w_i \left(1 + r_{i1}\right)\left(1 + r_{i2}\right) = \sum_{i=1}^{M} w_i \left(1 + r_{i1} + r_{i2} + r_{i1}r_{i2}\right) \\
&= 1 + \sum_{i=1}^{M} w_i r_{i1} + \sum_{i=1}^{M} w_i r_{i2} + \sum_{i=1}^{M} w_i r_{i1} r_{i2}
\end{aligned} \tag{10.12}
$$

The difference between the two gives rise to

$$W_{\mathrm{FW}} - W_{\mathrm{BH}} = \left(\sum_{i=1}^{M} w_i r_{i1}\right)\left(\sum_{i=1}^{M} w_i r_{i2}\right) - \sum_{i=1}^{M} w_i r_{i1} r_{i2} \tag{10.13}$$

We recognize that the weighted sums in the first term are simply the average returns in the first and second period, respectively, while the second term is the weighted average of the products of two sets of returns. In other words, we have

$$W_{\mathrm{FW}} - W_{\mathrm{BH}} = \overline{\mathbf{r}_1}\,\overline{\mathbf{r}_2} - \sum_{i=1}^{M} w_i r_{i1} r_{i2} = -\mathrm{cov}\left(\mathbf{r}_1, \mathbf{r}_2\right) \tag{10.14}$$

The difference between the terminal wealth of the two portfolios is related to the covariance of the two returns vectors in the two periods. We make several remarks about this result.

- Since covariance is positively proportional to cross-sectional correlation, Equation 10.14 proves that it is necessary to have cross-sectional reversal to have positive rebalancing alpha.
- When there are just two assets, the result simplifies to Equation 10.4. In fact, the result can be written in a sum with pairwise terms. We leave that as an exercise.

10.3 TWO-ASSET THREE-PERIOD CASE

The previous sections deal with cases with only two periods. When considering the difference between the terminal wealth of FW and BH portfolios, many terms cancel, because there is only limited compounding effect with just two periods. The analysis quickly gets more complicated as the number of periods increases. We now study the case with three periods and just two assets. A key insight is that the previous analysis can be extended once we combine two consecutive periods into a single period.

We start with the terminal wealth of the BH portfolio:

$$W_{\text{BH}} = w_1 \left(1+r_{11}\right)\left(1+r_{12}\right)\left(1+r_{13}\right) + w_2 \left(1+r_{21}\right)\left(1+r_{22}\right)\left(1+r_{23}\right) \tag{10.15}$$

To use the results from the case with two periods, we define compound returns during the second and third periods to have

$$1 + R_1^{2,3} = \left(1+r_{12}\right)\left(1+r_{13}\right)$$
$$1 + R_2^{2,3} = \left(1+r_{22}\right)\left(1+r_{23}\right) \tag{10.16}$$

Hence, $R_1^{2,3}$ is the return of the first asset during Periods 2 and 3, and $R_2^{2,3}$ is the return of the second asset during Periods 2 and 3. Equation 10.15 can be rewritten as

$$W_{\text{BH}} = w_1 \left(1+r_{11}\right)\left(1+R_1^{2,3}\right) + w_2 \left(1+r_{21}\right)\left(1+R_2^{2,3}\right) \tag{10.17}$$

Equation 10.17 resembles the terminal wealth of a BH portfolio for two periods. The terminal wealth of a corresponding FW portfolio is then

$$\tilde{W}_{FW} = \left(1 + w_1 r_{11} + w_2 r_{21}\right)\left(1 + w_1 R_1^{2,3} + w_2 R_2^{2,3}\right) \tag{10.18}$$

Note that $\tilde{W}_{FW}$ is not the terminal wealth of the FW portfolio, which is rebalanced after every period. Rather, this portfolio is only rebalanced after the first period but not after the second period.

The difference between the terminal wealth for the two portfolios, based on the result from Example 10.1, is

$$\tilde{W}_{FW} - W_{BH} = -w_1 w_2 \left(r_{11} - r_{21}\right)\left(R_1^{2,3} - R_2^{2,3}\right) \tag{10.19}$$

- Even though $\tilde{W}_{FW}$ is not the terminal wealth of the FW portfolio, it represents the terminal wealth of an actual portfolio that is rebalanced only once after the first period. Hence, our problem is reduced to a two-period problem, and Equation 10.19 determines whether rebalancing adds value for that single rebalance.

Now, we need to connect $\tilde{W}_{FW}$ to the actual terminal wealth W_{FW}. To do so, we note that

$$\left(1 + w_1 R_1^{2,3} + w_2 R_2^{2,3}\right) = w_1\left(1 + r_{12}\right)\left(1 + r_{13}\right) + w_2\left(1 + r_{22}\right)\left(1 + r_{23}\right) \tag{10.20}$$

This is just restating the fact that this is the terminal wealth of a portfolio for Periods 2 and 3. This terminal wealth is related to the terminal wealth of an FW portfolio for Periods 2 and 3 according to the same result given in Example 10.1. We have

$$\begin{aligned}
&w_1\left(1 + r_{12}\right)\left(1 + r_{13}\right) + w_2\left(1 + r_{22}\right)\left(1 + r_{23}\right) \\
&= \left(1 + w_1 r_{12} + w_2 r_{22}\right)\left(1 + w_1 r_{13} + w_2 r_{23}\right) + w_1 w_2\left(r_{12} - r_{22}\right)\left(r_{13} - r_{23}\right)
\end{aligned} \tag{10.21}$$

Substituting Equation 10.21 into Equation 10.18, we have

$$\tilde{W}_{FW} = W_{FW} + \left(1 + w_1 r_{11} + w_2 r_{21}\right) w_1 w_2\left(r_{12} - r_{22}\right)\left(r_{13} - r_{23}\right) \tag{10.22}$$

Finally, using Equation 10.19 in Equation 10.22 gives

$$W_{BH} - w_1 w_2 \left(r_{11} - r_{21} \right) \left(R_1^{2,3} - R_2^{2,3} \right) = W_{FW} + \left(1 + r_{FW,1} \right)$$
$$w_1 w_2 \left(r_{12} - r_{22} \right) \left(r_{13} - r_{23} \right) \tag{10.23}$$

We have used $r_{FW,1}$ to denote the return of the FW portfolio in the first Period 1. Therefore,

$$W_{FW} - W_{BH} = -w_1 w_2 \left(r_{11} - r_{21} \right) \left(R_1^{2,3} - R_2^{2,3} \right) - \left(1 + r_{FW,1} \right) w_1 w_2 \left(r_{12} - r_{22} \right) \left(r_{13} - r_{23} \right) \tag{10.24}$$

Equation 10.24 is again an exact result, which is remarkably similar to Equation 10.4 for two periods. We make the following remarks.

- Both terms consist of return differences between the two assets.

- The first term is proportional to differences in Period 1 and difference in the *combined* Periods 2 and 3.

- The second term is proportional to differences in Period 2 and difference in Period 3. However, there is an extra term that is related to portfolio return in Period 1.

In this case, whether rebalancing adds or detracts value is determined by two terms instead of one term, as in the case of two periods. Both terms can be interpreted as either momentum or reversal depending on the return differences between the two assets. In theory, each of these two terms can lead to either momentum or reversal. However, for rebalancing to add value, i.e. for $W_{FW} > W_{BH}$, at least one term on the right of Equation 10.24 must be positive. Supposing we have a long-only portfolio, then we must have reversal in at least one term. On the other hand, for rebalancing to detract value, for a long-only portfolio, we must have momentum in at least one term.

- Focusing on positive rebalancing alpha, it is clear now that the asset return must exhibit some degree of reversal over some periods.

- In Equation 10.24, the reversal occurs between Period 1 and combined Periods 2 and 3, or between Periods 2 and 3, or both.

Example 10.2: We consider a portfolio of two assets with 50% invested in each. The returns for three periods are displayed in Table 10.2. In Period 1, returns are −5% for Asset 1% and 5% for Asset 2; in Period 2, returns are −10% for Asset 1% and 0% for Asset 2; in Period 3, returns are 0% for Asset 1 and −5% for Asset 2. For the FW portfolio, the returns are 0% for Period 1, −5% for Period 2, and −2.5% for Period 3. So, its terminal wealth is $W_{FW} = (1+0\%)(1-5\%)(1-2.5\%) = 0.92625$. On the other hand, the terminal wealth of two assets invested on their own is 0.8550 and 0.9975, respectively. Hence, the terminal wealth of the BH portfolio is the average of the two due to the 50/50 weights. We have $W_{BH} = 0.5 \cdot 0.8550 + 0.5 \cdot 0.9975 = 0.92625$. In this case, miraculously, the FW portfolio and the BH portfolio have the same return; $W_{FW} - W_{BH} = 0$.

We use Equation 10.24 to illustrate this result. In Table 10.2, we also list the compound returns for Periods 2 and 3, which are −10% and −5%, respectively, for Assets 1 and 2. The first term in Equation 10.24, the rebalancing effect between Period 1 and combined Periods 2 and 3, is

$$-w_1w_2\left(r_{11}-r_{21}\right)\left(R_1^{2,3}-R_2^{2,3}\right) = -0.5^2 \cdot \left(-5\%-5\%\right)\left(-10\%+5\%\right)$$

$$= -0.00125 \tag{10.25}$$

On the other hand, the rebalancing effect between Periods 2 and 3 is

$$-\left(1+r_{FW,1}\right)w_1w_2\left(r_{12}-r_{22}\right)\left(r_{13}-r_{23}\right) = -0.5^2 \cdot \left(-10\%\right)\left(5\%\right)$$

$$= 0.00125 \tag{10.26}$$

The two effects perfectly offset each other, resulting in no overall rebalancing effect.

TABLE 10.2   Portfolio Weights and Period Returns of Two Assets for Three Periods

	Weight	Period 1	Period 2	Period 3	Wealth	Periods 2 + 3
Asset 1	50%	−5%	−10%	0%	0.8550	−10%
Asset 2	50%	5%	0%	−5%	0.9975	−5%
FW return		0.0%	−5.0%	−2.5%		

We note that in this case, if we had just rebalanced once after Period 1 and omitted the rebalance after Period 2, then the terminal wealth of the FW portfolio would have been higher by 0.00125.

There are two alternative versions of Equation 10.24 arising from different combinations of periods. For example, instead of combining Periods 2 and 3 together, we can combine Periods 1 and 2 together. Under this setup, we have

$$W_{FW} - W_{BH} = -w_1 w_2 \left(r_{13} - r_{23} \right) \left(R_1^{1,2} - R_2^{1,2} \right) - \left(1 + r_{FW,3} \right)$$

$$w_1 w_2 \left(r_{11} - r_{21} \right) \left(r_{12} - r_{22} \right) \tag{10.27}$$

We note that if we just rebalanced once after Period 2 and omitted the rebalance after Period 1, then the rebalancing effect would be given by the first term in Equation 10.27.

Alternatively, we can combine Periods 1 and 3 together. Then, the rebalancing effect will be

$$W_{FW} - W_{BH} = -w_1 w_2 \left(r_{12} - r_{22} \right) \left(R_1^{1,3} - R_2^{1,3} \right) - \left(1 + r_{FW,2} \right)$$

$$w_1 w_2 \left(r_{11} - r_{21} \right) \left(r_{13} - r_{23} \right) \tag{10.28}$$

Even though this calculation gives rise to the same result as Equations 10.24 and 10.27, we remark that it differs from them in one important aspect. Namely, the first term in Equation 10.28 does not represent a feasible rebalancing effect, since in reality, it is not feasible to combine Periods 1 and 3 together as a single BH period.

- The preceding discussion shows that between the BH and FW portfolios, there are other intermediate portfolios for which the rebalancing is only enacted occasionally. For an investment horizon of three periods, the FW portfolio is rebalanced twice. There are two intermediate portfolios rebalanced just once. In total, there are four portfolios depending on rebalancing frequency.

10.4 MULTIPLE-ASSET THREE-PERIOD CASE

It is quite straightforward to generalize the results in the previous section to portfolios with multiple assets with three investment periods by combining Equations 10.14 and 10.24. We have

$$W_{FW} - W_{BH} = -\text{cov}\left(\mathbf{r}_1, \mathbf{R}^{2,3}\right) - \left(1 + r_{FW,1}\right)\text{cov}\left(\mathbf{r}_2, \mathbf{r}_3\right) \qquad (10.29)$$

In Equation 10.29, $\mathbf{r}_1, \mathbf{r}_2, \mathbf{r}_3$ are vectors of asset returns for the three periods, and $\mathbf{R}^{2,3}$ is the vector of compound returns for both Periods 2 and 3.

Example 10.3: We extend the portfolio of two assets in Example 10.2 to a portfolio of three assets, with equal weights in all three. The returns for the third asset are added in Table 10.3. For the FW portfolio, the returns are 5% for Period 1, −5% for Period 2, and −0% for Period 3. So, its terminal wealth is $W_{FW} = (1 + 5\%)(1 - 5\%)(1) = 0.9975$. On the other hand, the terminal wealth of three assets invested on their own is 0.8550, 0.9975, and 1.147125, respectively. Hence, the terminal wealth of the BH portfolio is the average of the three due to equal weighting. We have $W_{BH} = 0.999875$. In this case, the FW portfolio underperforms the BH portfolio; $W_{FW} - W_{BH} = 0.9975 - 0.999875 = -0.002375$.

We can verify this result through Equation 10.29. The first term is the rebalancing effect between Period 1 and combined Periods 2 and 3. The covariance (population) between $(-5\%, 5\%, 15\%)$ and $(-10\%, -5\%, -0.25\%)$ is 0.00325. The second term is the rebalancing effect between Periods 2 and 3. The covariance (population) between $(-10\%, 0\%, -5\%)$ and $(0\%, -5\%, 5\%)$ multiplied by 1 plus 5% is −0.000875. Hence, we have $W_{FW} - W_{BH} = -0.00325 + 0.000875 = -0.002375$, matching the result by direct calculation.

TABLE 10.3 Returns and Weights of the Three-Asset Portfolio

	Weight	Period 1	Period 2	Period 3	Wealth	Periods 2 + 3
Asset 1	1/3	−5%	−10%	0%	0.8550	−10%
Asset 2	1/3	5%	0%	−5%	0.9975	−5%
Asset 3	1/3	15%	−5%	5%	1.1471	−0.25%
FW return		5%	−5%	0%		

The two alternative formulae corresponding to Equations 10.27 and 10.28 are

$$W_{FW} - W_{BH} = -\text{cov}\left(\mathbf{r}_3, \mathbf{R}^{1,2}\right) - \left(1 + r_{FW,3}\right)\text{cov}\left(\mathbf{r}_1, \mathbf{r}_2\right) \quad (10.30)$$

and

$$W_{FW} - W_{BH} = -\text{cov}\left(\mathbf{r}_2, \mathbf{R}^{1,3}\right) - \left(1 + r_{FW,2}\right)\text{cov}\left(\mathbf{r}_1, \mathbf{r}_3\right) \quad (10.31)$$

10.5 THE GENERAL CASE

We now derive the result for the general case of multiple assets and an investment horizon of multiple periods. The approach is similar to the one used in the case with three periods, and we extend the result to the general case by using it recursively.

The cumulative wealth of the FW portfolio is given by

$$W_{FW} = \left(1 + \mathbf{w}' \cdot \mathbf{r}_1\right)\left(1 + \mathbf{w}' \cdot \mathbf{r}_2\right)\cdots\left(1 + \mathbf{w}' \cdot \mathbf{r}_{N-1}\right)\left(1 + \mathbf{w}' \cdot \mathbf{r}_N\right) \quad (10.32)$$

We have used vector notations for the portfolio weights of the FW portfolio and returns in N different periods. Considering the last two periods of $N-1$ and N, we denote the compound returns of the two periods in a vector $\mathbf{R}^{N-1,N}$. Then by Equation 10.14, we have

$$\left(1 + \mathbf{w}' \cdot \mathbf{r}_{N-1}\right)\left(1 + \mathbf{w}' \cdot \mathbf{r}_N\right) = \left(1 + \mathbf{w}' \cdot \mathbf{R}^{N-1,N}\right) - \text{cov}\left(\mathbf{r}_{N-1}, \mathbf{r}_N\right) \quad (10.33)$$

Therefore,

$$\begin{aligned} W_{FW} = \left(1 + \mathbf{w}' \cdot \mathbf{r}_1\right)\cdots\left(1 + \mathbf{w}' \cdot \mathbf{r}_{N-2}\right)\left(1 + \mathbf{w}' \cdot \mathbf{R}^{N-1,N}\right) \\ - \left(1 + r_{FW,1}\right)\cdots\left(1 + r_{FW,N-2}\right)\text{cov}\left(\mathbf{r}_{N-1}, \mathbf{r}_N\right) \end{aligned} \quad (10.34)$$

We then consider the last two factors in the first term. We have

$$\left(1 + \mathbf{w}' \cdot \mathbf{r}_{N-2}\right)\left(1 + \mathbf{w}' \cdot \mathbf{R}^{N-1,N}\right) = \left(1 + \mathbf{w}' \cdot \mathbf{R}^{N-2,N}\right) - \text{cov}\left(\mathbf{r}_{N-2}, \mathbf{R}^{N-1,N}\right) \quad (10.35)$$

This leads to

$$W_{FW} = \left(1 + \mathbf{w}' \cdot \mathbf{r}_1\right) \cdots \left(1 + \mathbf{w}' \cdot \mathbf{r}_{N-3}\right)\left(1 + \mathbf{w}' \cdot \mathbf{R}^{N-2,N}\right)$$

$$-\left(1 + r_{FW,1}\right) \cdots \left(1 + r_{FW,N-3}\right) \text{cov}\left(\mathbf{r}_{N-2}, \mathbf{R}^{N-1,N}\right) \qquad (10.36)$$

$$-\left(1 + r_{FW,1}\right) \cdots \left(1 + r_{FW,N-2}\right) \text{cov}\left(\mathbf{r}_{N-1}, \mathbf{r}_N\right)$$

Continuing the procedure gives rise to

$$W_{FW} = \left(1 + \mathbf{w}' \cdot \mathbf{R}^{1,N}\right) - \text{cov}\left(\mathbf{r}_1, \mathbf{R}^{2,N}\right)$$

$$-\left(1 + r_{FW,1}\right)\text{cov}\left(\mathbf{r}_2, \mathbf{R}^{3,N}\right) - \cdots$$

$$\qquad (10.37)$$

$$-\left(1 + r_{FW,1}\right) \cdots \left(1 + r_{FW,N-3}\right)\text{cov}\left(\mathbf{r}_{N-2}, \mathbf{R}^{N-1,N}\right)$$

$$-\left(1 + r_{FW,1}\right) \cdots \left(1 + r_{FW,N-2}\right)\text{cov}\left(\mathbf{r}_{N-1}, \mathbf{r}_N\right)$$

We recognize the first term on the right-hand side as the wealth of the BH portfolio. Therefore, we obtain the general result

$$W_{FW} - W_{BH} = -\sum_{i=1}^{N-1}\left[\prod_{j=0}^{i-1}\left(1 + r_{FW,j}\right)\right]\text{cov}\left(\mathbf{r}_i, \mathbf{R}^{i+1,N}\right) \qquad (10.38)$$

We have defined $r_{FW,0} = 0$. Each term in Equation 10.38 consists of covariance between returns of one period and compound returns of all subsequent periods, and a coefficient that is the cumulative wealth of the FW portfolio prior to the particular period. In essence, each term accounts for the rebalancing benefit of one rebalancing. For example, the term $\text{cov}\left(\mathbf{r}_1, \mathbf{R}^{2,N}\right)$ is the benefit of rebalancing after the first period. In fact, if it was the only rebalance during the entire investment horizon, then this term represents the difference of the terminal wealth of this portfolio and the BH portfolio. The term $-\left(1 + r_{FW,1}\right)\text{cov}\left(\mathbf{r}_2, \mathbf{R}^{3,N}\right)$ measures the benefit of rebalancing of the second period. The term $\left(1 + r_{FW,1}\right)$ serves as a weight for this benefit, since the FW portfolio has grown to it at the beginning of the second period. If we only rebalance the portfolio after the first and second periods, then the rebalancing alpha equals the sum of these two terms.

We note that the coefficients are all positive.

- When the covariance is positive, it indicates momentum from one period to the rest of the investment horizon. When it is negative, it indicates reversal from one period to the rest of the investment horizon.

- If portfolio rebalancing adds value, i.e. $W_{FW} - W_{BH} > 0$, we conclude that at least one of the covariance terms must be negative. In other words, there exists reversal between one period and the rest of the investment horizon. On the other hand, if rebalancing detracts value, $W_{FW} - W_{BH} < 0$, we conclude that at least one of the covariance terms must be positive, implying that there exists momentum between one period and the rest of the investment horizon.

Example 10.4: We extend the portfolio of two assets in Example 10.3 to include a fourth period. The returns of the fourth period are 10%, 0%, and 0%, respectively, in Table 10.4. The return of the FW portfolio is 3.33% for the fourth period. The terminal wealth of the FW portfolio is $W_{FW} = (1 + 5\%)(1 - 5\%)(1)(1 + 3.33\%) = 1.03075$. On the other hand, the terminal wealth of three assets invested on their own is now 0.9405, 0.9975, and 1.147125, respectively. Hence, the terminal wealth of the BH portfolio is the average of the three due to equal weighting. We have $W_{BH} = 1.028375$. In this case, the FW portfolio outperforms the BH portfolio; $W_{FW} - W_{BH} = 1.030375 - 1.028375 = 0.002375$.

We now derive the rebalancing alpha using Equation 10.38. Explicitly, we have

$$W_{FW} - W_{BH} = -\text{cov}(\mathbf{r}_1, \mathbf{R}^{2,4}) - (1 + r_{FW,1})\text{cov}(\mathbf{r}_2, \mathbf{R}^{3,4})$$
$$-(1 + r_{FW,1})(1 + r_{FW,2})\text{cov}(\mathbf{r}_3, \mathbf{r}_4) \quad (10.39)$$

TABLE 10.4   Returns and Weights of the Three-Asset Portfolio for Four Periods

	Weight	$\mathbf{r}_1$	$\mathbf{r}_2$	$\mathbf{r}_3$	$\mathbf{r}_4$	Wealth	$\mathbf{R}^{2,4}$	$\mathbf{R}^{3,4}$
Asset 1	1/3	−5%	−10%	0%	10%	0.9405	−1%	10%
Asset 2	1/3	5%	0%	−5%	0%	0.9975	−5%	−5%
Asset 3	1/3	15%	−5%	5%	0%	1.1471	−0.25%	5.%
r_{FW}		5.0%	−5.0%	0.0%	3.3%			

The first term is $-\mathrm{cov}(\mathbf{r}_1, \mathbf{R}^{2,4}) = -0.00025$, indicating that the rebalancing after the first period detracts value. The second term is $-(1 + r_{\mathrm{FW},1})\mathrm{cov}(\mathbf{r}_2, \mathbf{R}^{3,4}) = 0.002625$, indicating that the rebalancing after the second period adds value. Finally, the last term is $-(1 + r_{\mathrm{FW},1})(1 + r_{\mathrm{FW},2})\mathrm{cov}(\mathbf{r}_3, \mathbf{r}_4) = 0$, since the covariance is zero, indicating that the rebalancing after the third period is neutral. Combining three terms together, we have $W_{\mathrm{FW}} - W_{\mathrm{BH}} = -0.00025 + 0.002625 + 0 = 0.002375$. This matches the result given by the direct computation.

Similarly to the previous result in Example 10.3, the general result in Equation 10.38 has many alternative forms. We note that if we rearrange the sequence of N periods, the terminal wealth of both BH and FW portfolios is unchanged. However, a rearrangement would lead to a different form of Equation 10.38. Specifically, if $(p_1, p_2, \cdots, p_N)$ is a permutation of $(1, 2, \cdots, N)$, then similarly to Equation 10.32, we have

$$W_{\mathrm{FW}} = \prod_{i=1}^{N}(1 + \mathbf{w}' \cdot \mathbf{r}_{p_i}) \qquad (10.40)$$

and the terminal wealth of the BH portfolio is just

$$W_{\mathrm{BH}} = (1 + \mathbf{w}' \cdot \mathbf{R}^{p_1, p_N}) \qquad (10.41)$$

Under this permutation of periods, applying the same procedure would lead to the rebalancing effect as

$$W_{\mathrm{FW}} - W_{\mathrm{BH}} = -\sum_{i=1}^{N-1}\left[\prod_{j=0}^{i-1}(1 + r_{\mathrm{FW},p_j})\right]\mathrm{cov}(\mathbf{r}_{p_i}, \mathbf{R}^{p_{i+1}, p_N}) \quad (10.42)$$

However, even though the rebalancing effect remains constant, its decomposition based on the terms on the right-hand side of Equation 10.42 will be different. As an example, we swap Period 1 and Period 4 in Example 10.4. The sequence of returns is displayed in Table 10.5. We have kept the same labels for the sequence of periods.

Of course, this rearrangement has no impact on the terminal wealth, so we still have $W_{\mathrm{FW}} - W_{\mathrm{BH}} = 1.030375 - 1.028375 = 0.002375$.

TABLE 10.5 Returns and Weights of the Three-Asset Portfolio for Four Periods

	Weight	r_1	r_2	r_3	r_4	Wealth	$R^{2,4}$	$R^{3,4}$
Asset 1	1/3	10%	−10%	0%	−5%	0.9405	−14.50%	−5.00%
Asset 2	1/3	0%	0%	−5%	5%	0.9975	−0.25%	−0.25%
Asset 3	1/3	0%	−5%	5%	15%	1.1471	14.71%	20.75%
r_{FW}		3.3%	−5.0%	0.0%	5%			

But the composition of the rebalancing effect is different. We list those three terms here:

$$-\text{cov}\left(\mathbf{r}_1, \mathbf{R}^{2,4}\right) = 0.0048292$$

$$-\left(1 + r_{FW,1}\right)\text{cov}\left(\mathbf{r}_2, \mathbf{R}^{3,4}\right) = -0.0008181 \qquad (10.43)$$

$$-\left(1 + r_{FW,1}\right)\left(1 + r_{FW,2}\right)\text{cov}\left(\mathbf{r}_3, \mathbf{r}_4\right) = -0.0016361$$

It is easy to show that the sum of three terms yields the same answer, $W_{FW} - W_{BH} = 0.002375$.

10.6 INCOMPLETE REBALANCE

The discussion in the previous section on the effect of portfolio rebalancing carried out in a specific period leads naturally to the problem of incomplete rebalance. In a BH portfolio, there is no rebalancing whatsoever. In an FW portfolio, rebalancing occurs at every period. Supposing there are N periods, in the BH portfolio there is zero rebalancing, and in the FW portfolio the number of rebalancings is $N - 1$. An incomplete rebalance would have the number of rebalancings between 1 and $N - 2$.

Given the number of rebalancings i, the number of choices to perform those rebalancings is given by C_{N-1}^i, which is the number of combination of choosing i elements from $N - 1$ elements. Hence, the total number of portfolio choices, including both the original FW and BH portfolios, is

$$\sum_{i=1}^{N-1} C_{N-1}^i = 2^{N-1} \qquad (10.44)$$

For instance (see Example 10.3), when there are three periods, besides the FW and BH portfolios, there are two other portfolios, in which the rebalancing is carried out just once: one after the first period and the other after the second period.

We can denote a particular portfolio by the period(s) when the rebalancing is performed. For example, if the rebalancing is only performed after the first period, the portfolio can be denoted as $p\{1\}$. If the rebalancing is performed after both the first and the third periods, the portfolio is $p\{1,3\}$. The FW portfolio is $p\{1,2,\cdots,N-1\}$.

Example 10.5: The cumulative wealth of a portfolio that is rebalanced once after the second period is

$$W_{p\{2\}} = \left(1 + \mathbf{w}' \cdot \mathbf{R}^{1,2}\right)\left(1 + \mathbf{w}' \cdot \mathbf{R}^{3,N}\right) \tag{10.45}$$

The rebalancing alpha of this single act is given by

$$W_{p\{2\}} - W_{\mathrm{BH}} = \left(1 + \mathbf{w}' \cdot \mathbf{R}^{1,2}\right)\left(1 + \mathbf{w}' \cdot \mathbf{R}^{3,N}\right) - \left(1 + \mathbf{w}' \cdot \mathbf{R}^{1,N}\right)$$

$$= -\mathrm{cov}\left(\mathbf{R}^{1,2}, \mathbf{R}^{3,N}\right) \tag{10.46}$$

Mathematically, this result is no different from the previous result that involved just two periods. In essence, we only have two periods too, albeit with potentially different length, and the returns are cumulative returns in these two periods.

In general, if the portfolio is denoted by $p\{i_1,\cdots,i_L\}$, there are L rebalancings. Then,

$$W_{p\{i_1,\cdots,i_L\}} = \left(1 + \mathbf{w}' \cdot \mathbf{R}^{1,i_1}\right)\left(1 + \mathbf{w}' \cdot \mathbf{R}^{i_1+1,i_2}\right)\cdots\left(1 + \mathbf{w}' \cdot \mathbf{R}^{i_L+1,N}\right) \tag{10.47}$$

The rebalancing effect of the given portfolio rebalancing can be calculated in the same way as given in Equation 10.38.

Among different choices of incomplete rebalancing schedules, we can explore schedules with maximum and minimum rebalancing alpha. One approach is to find maximum and minimum rebalancing alpha for a given number of rebalancings, starting with one rebalancing.

$$W_{p\{i\}} = \left(1 + \mathbf{w}' \cdot \mathbf{R}^{1,i}\right)\left(1 + \mathbf{w}' \cdot \mathbf{R}^{i+1,N}\right) \tag{10.48}$$

We are interested in the maximum and minimum of the terminal wealth $W_{p\{i\}}$.

Example 10.6: Given the returns of assets in Table 10.5, we have three possible schedules of one rebalancing, after Period 1, 2, or 3. In this case, recall that the terminal wealth of the BH portfolio is $W_{\text{BH}} = 1.028375$. From Table 10.6, we observe that the maximum wealth is achieved by rebalancing after Period 3, and the minimum is achieved by rebalancing after Period 1. Both $W_{p\{2\}}$ and $W_{p\{3\}}$ exceed the wealth of the BH portfolio, while $W_{p\{1\}}$ is below that of the BH portfolio.

If we choose to rebalance twice, there are three possibilities, listed in Table 10.7. The number in the first column indicates the time of the first rebalance, while the number in the row indicates the time of the second rebalance. Among the three, the terminal wealth $W_{p\{2,3\}}$ is the highest, and the terminal wealth $W_{p\{1,3\}}$ is the lowest. However, we note that $W_{p\{2,3\}} < W_{p\{3\}}$, which implies that the second rebalancing after Period 2 reduces the effect of rebalancing. This simple case also illustrates that sometimes less is more in terms of portfolio rebalancing and its efficacy. This is obviously true in hindsight. It is much more difficult in practice to know the future effect of rebalancing.

We can derive Equation 10.38 using a sequence of incomplete rebalances that links the FW and BH portfolios, which are the two ends of this sequence. Start with the BH portfolio, which we denote as $\{0,0,\cdots,0\}$, where 0 implies no rebalance. The next in the sequence is $\{1,0,\cdots,0\}$, where the rebalance only occurs after the first period. The next portfolio is $\{1,1,0,\cdots,0\}$, where the rebalance only occurs after both the first and the second periods. The sequence would continue

TABLE 10.6 The Terminal Wealth of Portfolios Based on Data in Table 10.4 and Rebalanced Just Once

$W_{p\{1\}}$	1.028125
$W_{p\{2\}}$	1.032472222
$W_{p\{3\}}$	1.033204167

TABLE 10.7 The Terminal Wealth of Portfolios Based on Data in Table 10.4 and Rebalanced Twice

	2	**3**
1	1.03075	1.029845833
2		1.032472222

as the rebalance is carried out for more periods until it reaches the FW portfolio.

We then compare the terminal wealth between two consecutive portfolios with incomplete rebalance.

$$W_{\mathrm{BH}} = W_{\{0,0,\cdots,0\}} = \mathbf{w}' \cdot \left(1 + \mathbf{R}^{1,N}\right) = \mathbf{w}' \cdot \left[\left(1 + \mathbf{r}_1\right)\left(1 + \mathbf{R}^{2,N}\right)\right] \qquad (10.49)$$

The terminal wealth of the portfolio rebalanced once after the first period is

$$W_{\{1,0,\cdots,0\}} = \left(1 + \mathbf{w}' \cdot \mathbf{r}_1\right)\left(1 + \mathbf{w}' \cdot \mathbf{R}^{2,N}\right) \qquad (10.50)$$

Then, the difference of the two is

$$
\begin{aligned}
W_{\{1,0,\cdots,0\}} - W_{\{0,0,\cdots,0\}} &= \left(1 + \mathbf{w}' \cdot \mathbf{r}_1\right)\left(1 + \mathbf{w}' \cdot \mathbf{R}^{2,N}\right) \\
&\quad - \mathbf{w}' \cdot \left[\left(1 + \mathbf{r}_1\right)\left(1 + \mathbf{R}^{2,N}\right)\right] \\
&= -\mathrm{cov}\left(\mathbf{r}_1, \mathbf{R}^{2,N}\right)
\end{aligned}
\qquad (10.51)
$$

This is exactly the effect of one rebalance. When the rebalance is done after both the first and the second period, the terminal wealth is

$$W_{\{1,1,\cdots,0\}} = \left(1 + \mathbf{w}' \cdot \mathbf{r}_1\right)\left(1 + \mathbf{w}' \cdot \mathbf{r}_2\right)\left(1 + \mathbf{w}' \cdot \mathbf{R}^{3,N}\right) \qquad (10.52)$$

The difference between Equations 10.52 and 10.50 is then

$$
\begin{aligned}
&W_{\{1,1,\cdots,0\}} - W_{\{1,0,\cdots,0\}} \\
&= \left(1 + \mathbf{w}' \cdot \mathbf{r}_1\right)\left(1 + \mathbf{w}' \cdot \mathbf{r}_2\right)\left(1 + \mathbf{w}' \cdot \mathbf{R}^{3,N}\right) - \left(1 + \mathbf{w}' \cdot \mathbf{r}_1\right)\left(1 + \mathbf{w}' \cdot \mathbf{R}^{2,N}\right) \\
&= \left(1 + \mathbf{w}' \cdot \mathbf{r}_1\right)\left\{\left(1 + \mathbf{w}' \cdot \mathbf{r}_2\right)\left(1 + \mathbf{w}' \cdot \mathbf{R}^{3,N}\right) - \mathbf{w}' \cdot \left[\left(1 + \mathbf{r}_2\right)\left(1 + \mathbf{R}^{3,N}\right)\right]\right\} \\
&= -\left(1 + \mathbf{w}' \cdot \mathbf{r}_1\right)\mathrm{cov}\left(\mathbf{r}_2, \mathbf{R}^{3,N}\right)
\end{aligned}
\qquad (10.53)
$$

This is the difference of terminal wealth between the portfolio rebalanced once (after the first period) and the portfolio rebalanced twice (after both the first and the second period). The first term in Equation 10.53 is simply the return of the first period, which is shared by both portfolios. The second covariance term is the cause of difference between the two portfolios.

Equation 10.53 can be generalized to any two consecutive portfolios in the sequence, whereas one portfolio has one more rebalance than the other portfolio. Summing over all the differences of two consecutive portfolios leads to Equations 10.37 and 10.38.

PROBLEMS

10.1. Prove that Equation 10.14 can be rewritten as

$$W_{FW} - W_{BH} = -\sum_{i<j} w_i w_j \left(r_{i1} - r_{j1} \right)\left(r_{i2} - r_{j2} \right) \tag{10.54}$$

10.2. Suppose we only rebalance a portfolio after the first and second periods and then keep it BH thereafter. Prove that the difference in terminal wealth between this portfolio and the true BH portfolio is

$$W - W_{BH} = -\text{cov}\left(\mathbf{r}_1, \mathbf{R}^{2,N} \right) - \left(1 + r_{FW,1} \right)\text{cov}\left(\mathbf{r}_2, \mathbf{R}^{3,N} \right) \tag{10.55}$$

10.3. Derive rebalancing alpha for a portfolio that is rebalanced after the first and the third period.

10.4. Derive terminal wealth in Tables 10.6 and 10.7.

CHAPTER 11

Risk and Return of Rebalancing Effects

ALMOST ALL PREVIOUS RESEARCH on portfolio rebalancing has focused exclusively on the return aspect of portfolio rebalancing. Portfolio rebalancing also has an impact on the volatility of portfolio returns. In this chapter, we shall use statistical analysis to derive both expectations and variances of the terminal wealth of fixed-weight (FW) and buy-and-hold (BH) portfolios. By focusing directly on terminal wealth, we are able to obtain analytic results, which shed light on the effects of portfolio rebalancing. In addition, we note that the results for terminal wealth are more relevant to true investment objectives than results for average returns. The objective is to analyze the risk-adjusted terminal wealth of both FW and BH portfolios.

The derivation of variance of terminal wealth is a laborious task even for serially independent returns. Therefore, for serially dependent returns, we only tackle a special case to examine the impact of both mean-reverting and trending returns on portfolio rebalancing.

11.1 TERMINAL WEALTH

This section restates the terminal wealth of the FW and BH portfolios. We also make some general assumptions about the return distributions of assets in these portfolios. Again, suppose we have M investable assets and an investment horizon of N periods. In the beginning, a portfolio is invested in the assets with weights $\mathbf{w} = (w_1, \ldots, w_M)'$, and the sum of the weights equals 1, i.e. $w_1 + w_2 + \cdots + w_M = 1$. For long-only portfolios, all weights are non-negative. When we consider long-short portfolios, we shall allow negative weights.

We denote the return of asset i in period n by r_{in} and assume that for individual assets, their return distributions, joint or marginal, while different for different assets, are the same for all time periods. The expected return of asset i is μ_i;

$$E(r_{in}) = \mu_i, \text{ for } i = 1, \cdots M, \text{ and } n = 1, \cdots, N \tag{11.1}$$

The covariance between assets i and j during any time period is σ_{ij};

$$E\left[(r_{in} - \mu_i)(r_{jn} - \mu_j)\right] = \sigma_{ij}, \text{ for } i, j = 1, \cdots, M, \text{ and } n = 1, \cdots, N \tag{11.2}$$

We have the covariance matrix of asset returns:

$$\Sigma = \begin{pmatrix} \sigma_{11} & \cdots & \sigma_{1M} \\ \vdots & \ddots & \vdots \\ \sigma_{M1} & \cdots & \sigma_{MM} \end{pmatrix} \tag{11.3}$$

For the FW portfolio, the portfolio return in period n is the weighted sum of individual returns with the fixed weights:

$$r_{FW,n} = w_1 r_{1n} + w_2 r_{2n} + \cdots + w_M r_{Mn} = \sum_{i=1}^{M} w_i r_{in} \tag{11.4}$$

The terminal wealth of the FW portfolio with an initial investment of unit value, after N periods, is the product

$$W_{FW} = \left(1 + \sum_{i=1}^{M} w_i r_{i1}\right) \cdots \left(1 + \sum_{i=1}^{M} w_i r_{iN}\right) = \prod_{n=1}^{N}\left(1 + \sum_{i=1}^{M} w_i r_{in}\right) \tag{11.5}$$

For the BH portfolio with the same initial weights, the terminal wealth is the weighted sum of the terminal wealth of individual assets. We have

$$W_{BH} = \sum_{i=1}^{M} w_i (1 + r_{i1}) \cdots (1 + r_{iN}) = \sum_{i=1}^{M} w_i \left[\prod_{n=1}^{N}(1 + r_{in})\right] \tag{11.6}$$

For the time being, we assume that returns from different periods are independent. There are zero serial correlations between asset returns of different time periods. Later in the chapter, we shall relax this condition for certain cases to study the effects of serial correlations. Note that we have not yet imposed any parametric form of return distribution.

11.2 EXPECTED TERMINAL WEALTH

Given the expected returns of all assets and the assumption of return independence, we now derive the expected terminal wealth of the FW and BH portfolios. For the FW portfolio, due to portfolio rebalancing, the expected return is the same weighted sum of expected returns of individual assets. We denote it by

$$\mu_p = w_1\mu_1 + w_2\mu_2 + \cdots + w_M\mu_M = \sum_{i=1}^{M} w_i\mu_i = \mathbf{w}' \cdot \boldsymbol{\mu} \qquad (11.7)$$

Similarly, the variance of the FW portfolio is defined as the portfolio variance:

$$\sigma_p^2 = \sum_{i,j=1}^{M} w_i w_j \sigma_{ij} = \mathbf{w}'\Sigma\mathbf{w} \qquad (11.8)$$

We shall use this definition in the next section when we derive the variance of terminal wealth.

Taking the expectation of Equation 11.5 and using serial independence of returns, we have

$$E\left(W_{\text{FW}}\right) = E\left[\left(1 + \sum_{i=1}^{M} w_i r_{i1}\right) \cdots \left(1 + \sum_{i=1}^{M} w_i r_{iN}\right)\right] = \left(1 + \mu_p\right)^N \qquad (11.9)$$

- Note that this equation resembles the definition of the sample geometric return of the FW portfolio defined in Chapter 3. However, the interpretation is quite different.

- The expected terminal wealth of the FW portfolio simply grows as a power function, with the growth rate being the weighted average of individual expected returns.

For the BH portfolio, we derive the portfolio terminal wealth using the terminal wealth of individual assets. Taking the expectation of Equation 11.6 yields

$$E(W_{BH}) = \sum_{i=1}^{M} w_i (1+\mu_i)^N \qquad (11.10)$$

- Similarly, this equation resembles the definition of the sample geometric return of the BH portfolio defined in Chapter 3. However, the interpretation is quite different.

- Since there is no portfolio rebalancing, the expected terminal wealth of the BH portfolio is just the weighted sum of the terminal wealth of individual assets.

11.2.1 Equal Expected Returns

First, we consider the case in which all individual assets have the same expected return. Then, it is easy to see that the expected terminal wealth in the two cases is also the same:

$$E(W_{FW}) = E(W_{BH}) = (1+\mu)^N \qquad (11.11)$$

We note that this result is much more general than these two particular portfolios. Indeed, when all assets have the same expected return for all time periods, as far as the expected return of the whole portfolio is concerned, it does not matter at all what the portfolio weights are.

11.2.2 General Case

What happens to the terminal wealth when the expected returns are not the same for all assets? Intuitively, one might expect portfolio rebalancing to become unfavorable. On average, the act of rebalancing would sell assets with higher expected returns and buy assets with lower expected returns. This implies that the expected terminal wealth of the FW portfolio would be lower than the expected terminal wealth of the BH portfolio.

It is useful to see the dynamics of rebalancing using a simple case of two assets and two periods. Starting with the same weights for both FW and BH approaches, the average terminal wealth at the end of the first period, denoted by subscript 1, is

$$E(W_{FW,1}) = E(W_{BH,1}) = w_1(1+\mu_1) + w_2(1+\mu_2) \tag{11.12}$$

The new "average" asset weights would be

$$w_1^* = \frac{w_1(1+\mu_1)}{w_1(1+\mu_1) + w_2(1+\mu_2)}$$
$$w_2^* = \frac{w_2(1+\mu_2)}{w_1(1+\mu_1) + w_2(1+\mu_2)} \tag{11.13}$$

Assuming $\mu_1 > \mu_2$ without loss of generality, we then have $w_1^* > w_1$ and $w_2^* < w_2$. In other words, the weight of Asset 1, which has a higher expected return, has increased, while the weight of Asset 2, the one with a lower expected return, has decreased. Since during the second period, the expected returns of the two assets are still μ_1 and μ_2, with $\mu_1 > \mu_2$, it would be more advantageous to hold on to the new weights than to revert to the old. In other words, the BH portfolio would be expected to have a higher return than the FW portfolio in the next period.

Mathematically, it is straightforward to show that at the end of the second period, the difference in expected terminal wealth is

$$E(W_{BH,2}) - E(W_{FW,2}) = w_1 w_2(\mu_1 - \mu_2)^2 \tag{11.14}$$

If both weights are positive, which is true for long-only portfolios, the right-hand side of Equation 11.14 will be positive, implying that the expected terminal wealth is higher for the BH portfolio than for the FW portfolio. Interestingly, though, if one of the weights is negative, the FW portfolio has higher expected terminal wealth instead. This is the case of long-short portfolios, a subject to which we shall return later.

For the general case of long-only portfolios, we recognize that by Jensen's inequality, we have

$$\left(1+\sum_{i=1}^{M} w_i\mu_i\right)^N \le \sum_{i=1}^{M} w_i(1+\mu_i)^N \tag{11.15}$$

Therefore, for long-only portfolios, we have

$$E(W_{FW}) \le E(W_{BH}) \tag{11.16}$$

We make several comments on this result.

- This result resembles the one regarding the geometric return of the BH portfolio, in that it is greater than the average of geometric returns of individual assets.

- It is apparent from our derivation that expected terminal wealth depends only on expected return of assets, not on volatilities or correlations of asset returns, provided that returns are serially independent. Therefore, diversification return or volatility effect, defined in terms of portfolio variance reduction, does not have any specific role in determining the expected terminal wealth for the FW and BH portfolios.

11.3 VARIANCE OF TERMINAL WEALTH

It appears that portfolio rebalancing is not desirable if the objective is to achieve higher expected terminal wealth. However, to render judgment on any portfolio approach, we must take into account the risks associated with the terminal wealth in addition to the expectation. Does portfolio rebalancing have any positive effects on the risks of terminal wealth? One suspects that the answer might be yes, and in fact, this could be one of the primary motivations for FW asset allocation portfolios used by disciplined investors. In this section, we derive analytic results on variances of terminal wealth to prove the case.

Being passive or more appropriately inactive, the asset weights of any BH portfolio could vary dramatically over time. In fact, given enough time, a BH portfolio will be dominated by assets with the highest realized returns, provided that some assets can persistently outperform other assets. Such concentrated portfolios will have higher risks even if the expected returns stay the same in the future. Put differently, portfolio diversification weakens gradually and persistently as the BH portfolio becomes more and more concentrated.

The variance of the terminal wealth of the BH portfolio depends on a sequence of portfolio risks that is constantly changing, whereas the variance of the terminal wealth of the FW portfolio depends on a single constant risk profile. In both cases, the variance of terminal wealth reflects the compounding effect of multiple portfolio risks. The question is: does portfolio rebalancing or staying with one risk profile lead to lower variance of the terminal wealth in the end?

To derive the variance of terminal wealth, we use the statistical formula for a random variable, $\mathrm{var}(x) = \mathrm{E}(x^2) - [\mathrm{E}(x)]^2$. Since we already derived $\mathrm{E}(W_{FW})$ and $\mathrm{E}(W_{BH})$ in the last section, what is required is the second moments $\mathrm{E}(W_{FW}^2)$ and $\mathrm{E}(W_{BH}^2)$.

For the FW portfolio, Square Equation 11.5 gives

$$W_{FW}^2 = \prod_{n=1}^{N}\left(1 + \sum_{i=1}^{M} w_i r_{in}\right)^2 \tag{11.17}$$

Taking expectation and using the assumption of inter-period independence, we have

$$\mathrm{E}(W_{FW}^2) = \prod_{n=1}^{N} \mathrm{E}\left[\left(1 + \sum_{i=1}^{M} w_i r_{in}\right)^2\right] \tag{11.18}$$

Using the identity $\mathrm{E}(x^2) = \mathrm{var}(x) + [\mathrm{E}(x)]^2$, we obtain

$$\mathrm{E}\left[\left(1 + \sum_{i=1}^{M} w_i r_{in}\right)^2\right] = \mathrm{var}\left(1 + \sum_{i=1}^{M} w_i r_{in}\right) + \left[\mathrm{E}\left(1 + \sum_{i=1}^{M} w_i r_{in}\right)\right]^2 \tag{11.19}$$

The first term is simply the variance of the fixed-weight portfolio σ_p^2, and the second term is 1 plus the expected return μ_p, both of which have been defined previously. Therefore,

$$\mathrm{E}\left[\left(1 + \sum_{i=1}^{M} w_i r_{in}\right)^2\right] = (1 + \mu_p)^2 + \sigma_p^2 \tag{11.20}$$

Inserting Equation 11.20 into Equation 11.18 yields

$$\mathrm{E}(W_{FW}^2) = \left[(1 + \mu_p)^2 + \sigma_p^2\right]^N \tag{11.21}$$

Equation 11.21 shows that the second moment of the terminal wealth of the FW portfolio is simply a power of the second moment of the wealth of each time period.

To derive $E\left(W_{\text{BH}}^2\right)$, we square Equation 11.6 to have

$$W_{\text{BH}}^2 = \left[\sum_{i=1}^{M} w_i \left(\prod_{n=1}^{N}(1+r_{in})\right)\right]^2 = \left[\sum_{i=1}^{M} w_i R_i\right]^2 = \sum_{i,j=1}^{M} w_i w_j R_i R_j \qquad (11.22)$$

We have defined the cumulative wealth of asset

$$R_i = \prod_{n=1}^{N}(1+r_{in}) \qquad (11.23)$$

We have

$$E\left(R_i R_j\right) = E\left[\prod_{n=1}^{N}(1+r_{in})(1+r_{jn})\right] = \left[(1+\mu_i)(1+\mu_j)+\sigma_{ij}\right]^N \qquad (11.24)$$

Therefore,

$$E\left(W_{\text{BH}}^2\right) = \sum_{i,j=1}^{M} w_i w_j \left[(1+\mu_i)(1+\mu_j)+\sigma_{ij}\right]^N \qquad (11.25)$$

For the BH portfolio, Equation 11.25 is a weighted sum of powers of covariance between different assets in a single time period.

Combining Equations 11.9 and 11.21, we obtain

$$\text{var}\left(W_{\text{FW}}\right) = \left[(1+\mu_p)^2+\sigma_p^2\right]^N-(1+\mu_p)^{2N} \qquad (11.26)$$

and combining Equations 11.10 and 11.25 gives

$$\text{var}\left(W_{\text{BH}}\right) = \sum_{i,j=1}^{M} w_i w_j \left[(1+\mu_i)(1+\mu_j)+\sigma_{ij}\right]^N - \left[\sum_{i=1}^{M} w_i(1+\mu_i)^N\right]^2 \qquad (11.27)$$

To relate the variance of terminal wealth to the variances and covariances of asset returns, we expand the first term in both and then cancel the second term associated with expected terminal wealth. From Equation 11.26, we have

$$\text{var}\left(W_{\text{FW}}\right) = \sum_{n=1}^{N} C_N^n \left(1+\mu_p\right)^{2(N-n)} \sigma_p^{2n} \qquad (11.28)$$

Equation 11.28 shows that the variance of the terminal wealth of the FW portfolio is a binomial sum of the variances of different time horizons (n) multiplied by expected wealth cumulated during the remaining (or proceeding) time horizon $(N-n)$, with $C_N^n = N!/\left[n!(N-n)!\right]$. For instance, take $n=1$; the variance for a single period is of course σ_p^2. The cumulative wealth squared during the remaining $(N-1)$ periods is $(1+\mu_p)^{2(N-1)}$. This wealth, which has grown according to the rate μ_p, is now subject to the volatility of the "last" period. We also note from Equation 11.28 that the variance is higher for higher portfolio volatility, and it is also higher for higher expected return.

We treat Equation 11.27 similarly by expanding the first term and then cancelling the second term. The result is

$$\text{var}\left(W_{\text{BH}}\right) = \sum_{n=1}^{N} C_N^n \sum_{i,j=1}^{M} w_i w_j \left[\left(1+\mu_i\right)\left(1+\mu_j\right)\right]^{N-n} \sigma_{ij}^n \qquad (11.29)$$

The variance of the terminal wealth of the BH portfolio is thus a binomial sum of the modified covariance between different assets, in which powers of covariance of time horizon n are modified by the wealth of assets accumulated in time horizon $(N-n)$.

11.4 COMPARISON OF TWO VARIANCES

Unlike the expected terminal wealth, a general result concerning the relative magnitude of the two variances does not seem to be available. Nevertheless, we shall use some special cases to gain insight into the effects of portfolio rebalancing on the variance of terminal wealth.

The first case is when all assets have the same expected return. It is known already that the expected terminal wealth is the same for the FW and BH portfolios. Equation 11.28 for the variance of the terminal wealth of the FW portfolio simplifies to

$$\text{var}\left(W_{\text{FW}}\right) = \left(1+\mu\right)^{2N} \sum_{n=1}^{N} \left[C_N^n \left(1+\mu\right)^{-2n} \sigma_p^{2n}\right] \qquad (11.30)$$

and the variance of the terminal wealth of the BH portfolio simplifies to

$$\text{var}\left(W_{\text{BH}}\right) = \left(1+\mu\right)^{2N} \sum_{n=1}^{N} \left[C_N^n \left(1+\mu\right)^{-2n} \sum_{i,j=1}^{M} w_i w_j \sigma_{ij}^n \right] \tag{11.31}$$

The two summations now have the same coefficients and the same powers of expected returns. The only difference is in the variance and covariance terms. To compare Equations 11.30 and 11.31, we compare the volatility term σ_p^{2n} and $\sum_{i,j=1}^{M} w_i w_j \sigma_{ij}^n$.

When $n=1$, the two terms are identical according to the definition of portfolio variance of the FW portfolio. When $n>1$, we can prove that for long-only portfolios and non-negative asset correlations, the following inequality holds:

$$\sigma_p^{2n} \leq \sum_{i,j=1}^{M} w_i w_j \sigma_{ij}^n \tag{11.32}$$

We rewrite the left-hand side in terms of portfolio weights and covariances:

$$\left(\sum_{i,j=1}^{M} w_i w_j \sigma_{ij} \right)^n \leq \sum_{i,j=1}^{M} w_i w_j \sigma_{ij}^n \tag{11.33}$$

In other words, the power of the weighted average of covariances is less than the weighted average of covariances powered. This inequality is true because of Jensen's inequality on a power function $f(x) = x^n$, $x > 0$. When $n > 1$, the power function is convex. We also have

$$\sum_{i,j=1}^{M} w_i w_j = \left(\sum_{i=1}^{M} w_i \right)^2 = 1 \tag{11.34}$$

- The equality in Equation 11.33 occurs when all covariances are the same, which is only true when all assets have the same volatility and all pairwise correlations equal 1, i.e. all assets are identical.

- In fact, the requirement that $x > 0$ is not needed if n is even. In that case, the function $f(x) = x^n$ is convex for all real x. Therefore, the inequality in Equation 11.33 holds for arbitrary covariances. This makes it highly likely that the variance of the FW portfolio will still be less than that of the BH portfolio.

Combining Equations 11.33, 11.30, and 11.31 gives

$$\text{var}\left(W_{\text{FW}}\right) \leq \text{var}\left(W_{\text{BH}}\right) \tag{11.35}$$

Therefore, the variance of the terminal wealth of the FW portfolio is always less than or equal to the variance of the terminal wealth of the BH portfolio when all pairwise correlations and portfolio weights are positive.

When the expected return is the same for all assets, the effect of portfolio rebalancing is variance reduction. Since the expected returns are the same for the two portfolios, we conclude that the FW portfolio has a better risk-adjusted terminal wealth than the BH portfolio.

- With the same expected return, the BH portfolio does not have any persistent drift in portfolio weights. However, its portfolio weights at any given period are still random, even though their averages are the original weights. It is possible that this additional randomness in its weights causes the additional variance of the terminal wealth of the BH portfolio.

- We note that the conditions under which this inequality holds are not as restricted as it seems at first. For asset allocation portfolios, it is highly likely that the expected returns of different asset classes, such as stocks, bonds, and commodities, are different because of the vast differences in their risk levels. Within the same asset class, however, it is reasonable to expect similar returns from different securities. In addition, correlations within the same asset classes are almost always positive. For example, in an equity universe, pairwise correlations are mostly always positive. The expected returns or realized returns are not necessarily the same. However, all equity returns reflect certain common risk factors over the long term, and so their average returns should not be very dissimilar. In fact, many empirical tests of the capital asset pricing model (CAPM) found the capital market line within the equity universe to be quite flat, indicating similar returns for stocks with different levels of risk.

- Furthermore, we note that when asset return correlations are negative, the FW portfolio typically has a lower variance. This is because the inequality in Equation 11.33 is true for all possible correlations if n is even. This guarantees that half of the terms in the summation (Equation 11.30) will be lower than the corresponding terms in the summation (Equation 11.31) and with the leading term ($n = 2$) in favor of the FW portfolio.

Example 11.1 We have portfolios based on 10 assets whose annual expected returns are all 8%. For simplicity, we further assume that their annual return volatility is 20% and their pairwise correlations are identical. We study three cases with three different correlation assumptions. The initial portfolio is equally weighted with 10% in each of the 10 assets.

Figure 11.1 displays the mean of terminal wealth of both FW and BH portfolios for investment horizons ranging from 1 to 20 years. Since the expected return is identical for all assets, the expected terminal wealth is the same for both FW and BH portfolios.

The volatility of the terminal wealth would also grow over time. We first assume that the pairwise correlations are zero. Figure 11.2 shows the volatilities of the terminal wealth for the FW and BH portfolios over different horizons. Both grow exponentially from zero at the beginning, and as predicted by our analysis, the volatility of the BH portfolio is higher than the volatility of the FW portfolio for all horizons. It is more relevant to compare the return/risk ratio of terminal wealth. We use expected terminal wealth less 1 as the return term. In Figure 11.3, we plot the ratio for the FW and BH portfolios. As seen from the graph, the ratios are increasing functions of the horizon, and the ratio is always higher for the FW portfolio.

The next case we consider is when all pairwise correlations are 0.4, while all other assumptions remain the same. In this case, the expected terminal wealth stays the same as in Figure 11.1, while the volatilities of terminal wealth increase due to high correlations. Figure 11.4 displays the volatilities of the two portfolios as the investment horizon varies from 1 to 20 years. It can be seen that while the line for the FW portfolio is always below the line for the BH portfolio; the two are very close to each other. It could be the case that when pairwise correlation is at this level, all assets move much more in tandem; thus, portfolio rebalancing makes little difference.

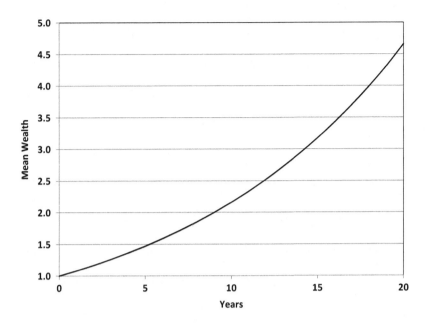

FIGURE 11.1   Expected terminal wealth.

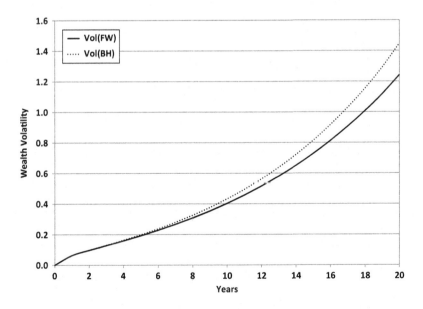

FIGURE 11.2   Volatility of terminal wealth for the FW and BH portfolios when correlations are zero.

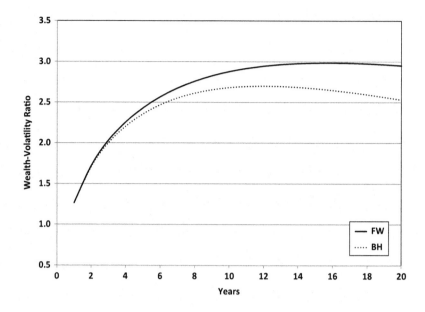

FIGURE 11.3   Wealth-volatility ratio of the FW and BH portfolios when correlations are zero.

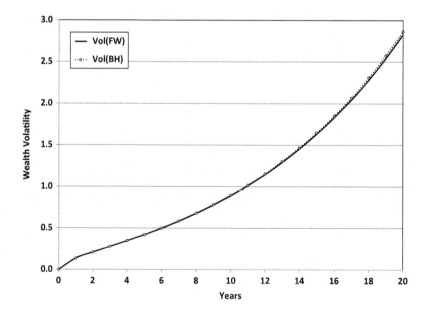

FIGURE 11.4   Volatility of terminal wealth for the FW and BH portfolios when correlations are 0.4.

In addition, the volatilities of terminal wealth are much higher than the volatilities in the case where the correlations are zero. In Figure 11.5, we plot the return/risk ratio for the FW and BH portfolios. The ratios are barely distinguishable. The ratio is close to 3 for the FW portfolio when the correlation is zero, as seen in Figure 11.3, but it drops to 1.3 for the higher-correlation case in Figure 11.5.

The third case we consider is when the pairwise correlation is −0.1. This case of negative correlation is intriguing, because it is not covered in our proof that the volatility of terminal wealth is lower for the FW portfolio than for the BH portfolio. It turns out that this is still true. Figure 11.6 displays the volatilities of the two portfolios as the investment horizon lengthens from 1 to 20 years. It can be seen that not only is the volatility of terminal wealth lower for the FW portfolio than for the BH portfolio, but the difference between the two is much greater than those in the previous case when correlations are zero. In Figure 11.7, we plot the ratio of expected terminal wealth to its standard deviation. Portfolio rebalancing has a far greater effect in this case. The mean volatility ratio of terminal wealth is much higher for the FW portfolio than for the BH portfolio. It appears that even though a mathematical proof has eluded us so far, the risk

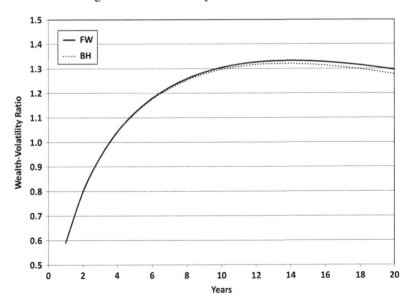

FIGURE 11.5 Wealth-volatility ratio of the FW and BH portfolios when correlations are 0.4.

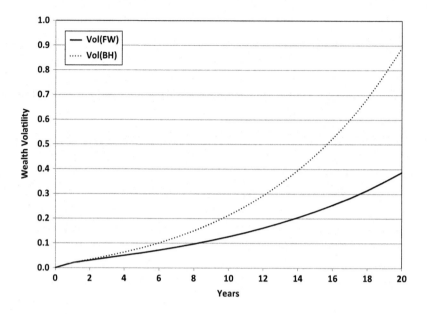

FIGURE 11.6 Volatility of terminal wealth for the FW and BH portfolios when correlations are −0.1.

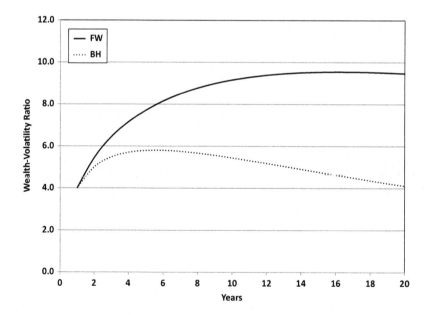

FIGURE 11.7 Wealth-volatility ratio of the FW and BH portfolios when correlations are −0.1.

reduction by the FW portfolio is even stronger when the correlations are negative. This is probably due to the fact that with negatively correlated assets, portfolio drift could be more substantial, thus allowing portfolio rebalancing to be more impactful.

11.5 A GENERAL TWO-ASSET CASE

To gain further insights into the effects of portfolio rebalancing, especially when the expected returns are different for different assets, we consider a case with two assets, one of which is risk-free. We denote the first asset as the risky asset with subscript 1 and the second asset as the risk-free asset with subscript 0. For the risk-free asset, we assume zero return volatility, i.e. $\sigma_0 = 0$.

Then, according to Equations 11.9 and 11.10, we have

$$
\begin{aligned}
E(W_{FW}) &= (1 + w_1\mu_1 + w_0\mu_0)^N \\
E(W_{BH}) &= w_1(1+\mu_1)^N + w_0(1+\mu_0)^N
\end{aligned}
\tag{11.36}
$$

And assuming $\mu_1 > \mu_0$, we have $E(W_{BH}) > E(W_{FW})$.

Since one asset is risk-free, the covariance between the two assets is also zero, i.e. $\sigma_{01} = 0$. The variances of terminal wealth for both the FW and the BH portfolio would depend only on the variance of the risky asset. From Equation 11.28 and 11.29, we have

$$
\text{var}(W_{FW}) = \sum_{n=1}^{N}\left[C_N^n (1 + w_1\mu_1 + w_0\mu_0)^{2N-2n} (w_1\sigma_1)^{2n} \right]
\tag{11.37}
$$

$$
\text{var}(W_{BH}) = w_1^2 \sum_{n=1}^{N}\left[C_N^n (1+\mu_1)^{2N-2n} (\sigma_1)^{2n} \right]
\tag{11.38}
$$

It is easy to prove that for long-only portfolios, the variance of the BH portfolio is always greater than the variance of the FW portfolio. Since $\mu_1 > \mu_0$, we have $\mu_1 > w_1\mu_1 + w_0\mu_0$ when both weights are positive, i.e. the expected return of the FW portfolio is less than that of the risky asset. It follows that

$$
\text{var}(W_{FW}) < \sum_{n=1}^{N}\left[C_N^n (1+\mu_1)^{2N-2n} (w_1\sigma_1)^{2n} \right]
\tag{11.39}
$$

Because $0 < w_1 < 1$, we have $w_1^{2n} \leq w_1^2$ for $n > 1$. Then,

$$\text{var}\left(W_{\text{FW}}\right) < w_1^2 \sum_{n=1}^{N}\left[C_N^n \left(1+\mu_1\right)^{2N-2n}\left(\sigma_1\right)^{2n}\right] = \text{var}\left(W_{\text{BH}}\right) \quad (11.40)$$

Thus, in this case, portfolio rebalancing leads to a lower expected terminal wealth but also to a lower variance. Unlike the cases in the previous section, the expected returns are different in the present case. Still, the variance reduction remains true for the FW portfolio. It thus appears that variance reduction could be a general feature of portfolio rebalancing, except perhaps for some extremely unlikely combinations of portfolio weights and return covariances.

To evaluate the combined effects on both return and risk in terms of terminal wealth, we shall consider several return/risk ratios. One such ratio is defined as the expected terminal wealth less unity divided by the standard deviation of terminal wealth. Another ratio, which is similar to the Sharpe ratio of asset returns, is defined by expected terminal wealth in excess of the terminal wealth of the risk-free asset, divided by the standard deviation of terminal wealth.

We now consider a return/risk ratio for the two-asset case. Since we have designated a risk-free asset in the present case, we shall consider a wealth Sharpe ratio as

$$\text{SR}_W = \frac{E\left(W\right)-\left(1+\mu_0\right)^N}{\text{std}\left(W\right)} \quad (11.41)$$

When N is large, this ratio is dominated by the ratio $E\left(W\right)/\text{std}\left(W\right)$. It can be proved that (see Appendix 11.A)

$$\frac{E\left(W_{\text{FW}}\right)}{\text{std}\left(W_{\text{FW}}\right)} > \frac{E\left(W_{\text{BH}}\right)}{\text{std}\left(W_{\text{BH}}\right)} \quad (11.42)$$

Therefore, the risk-adjusted terminal wealth is higher for the FW portfolio than for the BH portfolio.

It is highly likely that the wealth Sharpe ratio is also higher for the FW portfolio. We illustrate these results with two numerical examples. In both, the risk-free asset has an expected return of 1% per year.* In the first exam-

---

* While the return assumption of risk-free asset is rather arbitrary for our analysis, we assign it a rather low level, acknowledging the low–interest rate environment at the time of writing.

ple, the risky asset has an expected return of 2.5% and volatility of 5%, similar to that of investment-grade bonds. In both examples, the initial portfolio weights are 50% in the risky asset and 50% in the risk-free asset.

Figure 11.8 shows the expected terminal wealth and volatility of this example. Assuming expected returns of 1% and 2.5%, the expected terminal wealth grows slowly to about 1.4 in 20 years, with the BH portfolio having a higher expected terminal wealth. The volatility is displayed as well, with its scale on the right-hand axis. Since the risky asset has rather low volatility, the volatilities of the terminal wealth grow with time but remain low compared with its expectation. As we have proved, the FW portfolio always has a lower volatility of terminal wealth.

Figure 11.9 plots the wealth Sharpe ratio for the two portfolios. As the horizon lengthens, the ratios increase from an initial value of 0.3 to above 1. The graph demonstrates that the ratio is indeed always higher for the FW portfolio than it is for the BH portfolio, implying that the FW portfolio is preferred on a risk-adjusted basis.

In the second example, the risky asset has an expected return of 8.0% and volatility of 20%, serving as a proxy for equities. In this case, the expected terminal wealth will grow more quickly, but so does its

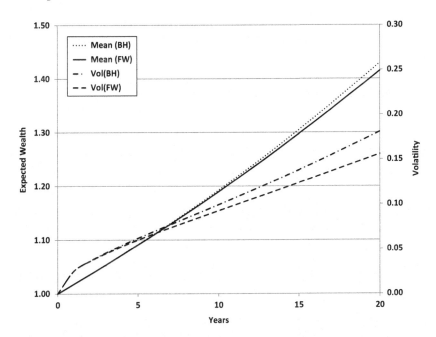

FIGURE 11.8 Expected terminal wealth and volatility of the FW and BH portfolios of two assets.

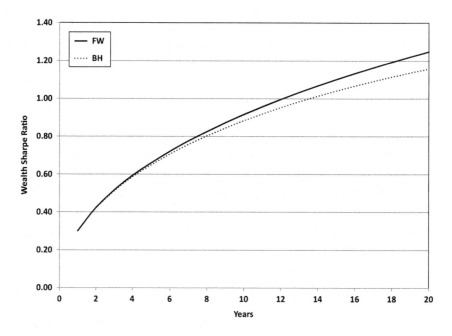

FIGURE 11.9   Wealth Sharpe ratio of the FW and BH portfolios of two assets.

volatility. Figure 11.10 shows the expected terminal wealth and volatility. The expected terminal wealth grows from 1 to 2.4 and 2.9 for the FW and BH portfolios, respectively, and it is always higher for the BH portfolio. The volatility of terminal wealth grows from 0 to 1.2 and 2.3 for the FW and BH portfolios, respectively. In other words, the volatility of the BH portfolio is almost twice as high as the volatility of the FW portfolio for an investment horizon of 20 years. An important reason is that as the horizon lengthens, the BH portfolio has a higher allocation to the risky asset.

Figure 11.11 plots the wealth Sharpe ratio of terminal wealth for the two portfolios. The ratio is always better for the FW portfolio. As the horizon lengthens, the wealth Sharpe ratio of the BH portfolio hits a plateau near 0.8, while the wealth Sharpe ratio of the FW portfolio continues to increase. Again, the graph demonstrates that the FW portfolio is preferred on a risk-adjusted basis.

11.6 THE IMPACT OF SERIAL CORRELATIONS

So far, we have derived our results regarding portfolio rebalancing under the assumption that returns from different time periods are independent. This assumption has made the analysis more tractable. We now relax this

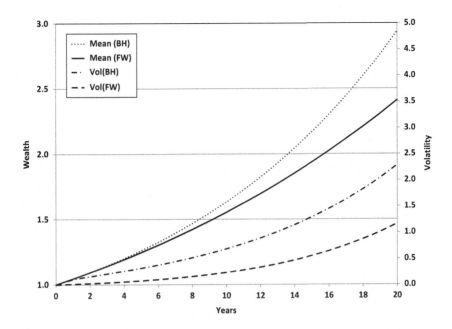

FIGURE 11.10   Expected terminal wealth and volatility of the FW and BH portfolios of the second set of two assets.

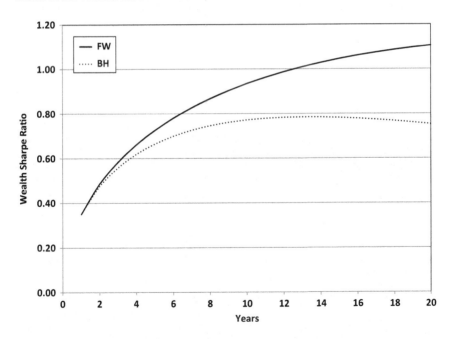

FIGURE 11.11   Wealth Sharpe ratio of the FW and BH portfolios of the second set of two assets.

assumption by considering cases in which asset returns have serial, auto-, and cross-correlations.

The two phenomena associated with serial correlations are return momentum and reversal. We are concerned with neither the existence nor the justification of such phenomena but rather, their implications for portfolio rebalancing. A full analysis of expected terminal wealth and its variance for the general case is possible but quite cumbersome. Furthermore, we may not gain much insight from such a laborious analysis. Instead, we consider a simple case with just two time periods, for which serial correlations of one period lag are incorporated. The impact of these serial correlations on both expected terminal wealth and its variance can be derived analytically, and the results are relatively intuitive. The detail of the derivation is in Appendix 11.B. We summarize the results here and then illustrate them with a specific example.

Assume that the lag-one covariance between the return of asset i in Period 1 and the return of asset j in Period 2 is

$$E\left[\left(r_{i1} - \mu_i\right)\left(r_{j2} - \mu_j\right)\right] = \sigma_{ij}^{1,2}, i, j = 1, \cdots, M \tag{11.43}$$

We have used superscript 1 and 2 to denote the covariances of different time periods. In matrix form, we write

$$\Sigma^{1,2} = \begin{pmatrix} \sigma_{11}^{1,2} & \cdots & \sigma_{1M}^{1,2} \\ \vdots & \ddots & \vdots \\ \sigma_{M1}^{1,2} & \cdots & \sigma_{MM}^{1,2} \end{pmatrix} \tag{11.44}$$

The expected terminal wealth of the FW portfolio after two time periods is

$$E\left(W_{\mathrm{FW}}\right) = \left(1 + \sum_{i=1}^{M} w_i \mu_i\right)^2 + \sum_{i,j=1}^{M} w_i w_j \sigma_{ij}^{1,2} = \left(1 + \mu_p\right)^2 + \vec{w}'\Sigma^{1,2}\vec{w} \tag{11.45}$$

Note the first term is the expected value of the terminal wealth absent any serial correlations. The second term, consisting of the weighted sum of serial covariances between different assets, is the additional term.

For the BH portfolio, the expected terminal wealth is given by

$$E\left(W_{\mathrm{BH}}\right) = \sum_{i=1}^{M} w_i \left(1 + \mu_i\right)^2 + \sum_{i=1}^{M} w_i \sigma_{ii}^{1,2} \tag{11.46}$$

Similarly, the first term is the expected value of the terminal wealth absent any serial correlations. The second term consists of the weighted sum of serial covariances between the same assets. There is no interaction between different asset returns from different time periods. Both expected terminal wealth values are functions of expected returns as well as the serial correlations of returns.

The variance of terminal wealth for the FW portfolio is given by

$$\operatorname{var}(W_{\mathrm{FW}}) = 2(1+\mu_p)^2 \sigma_p^2 + \sigma_p^4 + 2(1+\mu_p)^2 (\vec{w}' \Sigma^{1,2} \vec{w}) + \left[(\vec{w}' \Sigma^{1,2} \vec{w})\right]^2 \quad (11.47)$$

The first two terms give the variance of terminal wealth absent any serial correlations. The last two terms reflect the additional influences of serial correlations. Note that the two parts are of similar form.

For the BH portfolio, the variance of terminal wealth is

$$\operatorname{var}(W_{\mathrm{BH}}) = \sum_{i,j=1}^{M} w_i w_j \left[2(1+\mu_i)(1+\mu_j)\sigma_{ij} + \sigma_{ij}^2 \right]$$

$$+ \sum_{i,j=1}^{M} w_i w_j \left[2(1+\mu_i)(1+\mu_j)\sigma_{ij}^{1,2} + \sigma_{ij}^{1,2}\sigma_{ji}^{1,2} \right] \quad (11.48)$$

Similarly, the first summation gives the variance of terminal wealth absent any serial correlations, and the second summation reflects the additional influences of serial correlations. Note again that the two parts are of similar form. It remains to be seen whether the symmetry displayed in both Equations 11.47 and 11.48 is preserved when serial covariances with higher lags are included.

These results depend on expected returns, contemporaneous covariances, and serial covariances of all assets. They can be used to compare the performance of the FW and BH portfolios with serial correlations under general conditions. To gain insight into the effect of serial correlation, however, we focus on a special case that has been examined before: the case with two assets—one risky and the other risk-free.

As before, we denote the weights by w_1, w_0 and expected returns by μ_1, μ_0 for the risky asset and the risk-free asset, respectively, and the volatility of the risky asset as σ_1. In addition, we assume that the serial correlation of the risky asset with one lag is ρ_1. Then, according to Equation 11.45, the expected terminal wealth of the FW portfolio is

$$E\left(W_{\text{FW}}\right) = \left(1 + w_1\mu_1 + w_0\mu_0\right)^2 + \rho_1 w_1^2 \sigma_1^2 \qquad (11.49)$$

And the expected terminal wealth of the BH portfolio, according to Equation 11.46, is

$$E\left(W_{\text{BH}}\right) = w_1\left(1 + \mu_1\right)^2 + w_0\left(1 + \mu_0\right)^2 + \rho_1 w_1 \sigma_1^2 \qquad (11.50)$$

We make several remarks.

- The effect of the serial correlation on the expected terminal wealth in both portfolios, with positive weight w_1, has the same sign as the serial correlation. Namely, if the serial correlation ρ_1 is positive, which indicates momentum or trending in the returns of the risky asset, the expected terminal wealth of both increases. On the other hand, if the serial correlation is negative, which implies mean reversion in the risky asset, the expected terminal wealth of both decreases.

- However, the magnitudes of changes are different for the two portfolios. Relatively, the change is smaller for the FW portfolio, because we have $w_1^2 < w_1$ in the terms involving serial correlation. Hence, it is true that mean reversion favors the FW portfolio compared with the BH portfolio, because the negative impact on the FW portfolio is smaller.

- Conversely, it is also true that trending returns favor the BH portfolio compared with the FW portfolio.

From the previous discussion, we know that the expected wealth of the FW portfolio is less than that of the BH portfolio absent serial correlation. Hence, a natural question is whether or not the effect of negative serial correlation could be large enough to close the gap between the two. We address this question by considering the difference between the two expected values of Equations 11.49 and 11.50. We have

$$E\left(W_{\text{FW}}\right) - E\left(W_{\text{BH}}\right) = -w_0 w_1 \sigma_1^2 \left[\left(\frac{\mu_1 - \mu_0}{\sigma_1}\right)^2 + \rho_1\right] \qquad (11.51)$$

Therefore, if the serial correlation is less than the threshold given by

$$\rho_1 < -\left(\frac{\mu_1 - \mu_0}{\sigma_1}\right)^2 \tag{11.52}$$

we have $E(W_{FW}) > E(W_{BH})$. It can be recognized that the right-hand side of Equation 11.52 is the square of the Sharpe ratio for the risky asset. Hence, the necessary condition for the FW portfolio to have higher expected terminal wealth than the BH portfolio is that the serial correlation of the risky asset be less than the negative of its Sharpe ratio squared.

- The inequality in Equation 11.52 can only be the case if the Sharpe ratio in question is less than unity.

- In practice, this requirement does not present a problem over the long run, since long-term Sharpe ratios of all risky assets have been far less than unity. For instance, the Sharpe ratios of stocks and bonds have been between 0.3 and 0.4.

- Note that the condition in Equation 11.52 is independent of portfolio weights and the volatility of the risky asset.

- Numerically, when the Sharpe ratio is 0.3, then the condition in Equation 11.52 is $\rho_1 < -0.09$. If the Sharpe ratio is 0.2, then the condition is $\rho_1 < -0.04$. We comment that these conditions do not appear too stringent, especially when the Sharpe ratio of the risky asset is modest. It is further noted that for cases with more than two assets, the FW portfolio with portfolio rebalancing will benefit more if all assets exhibit some degree of negative serial auto- and cross-correlations.

- If the return of the risky asset is trending instead of mean-reverting, the gap in the expected terminal wealth between the FW and the BH portfolio will widen.

The serial correlation also affects the variances of the terminal wealth. For the case under consideration, we have, for the FW portfolio,

$$\text{var}(W_{FW}) = w_1^4 \sigma_1^4 (1 + \rho_1^2) + 2w_1^2 \sigma_1^2 (1 + \rho_1)(1 + w_1\mu_1 + w_0\mu_0)^2 \tag{11.53}$$

For the BH portfolio, the variance of terminal wealth is

$$\text{var}\left(W_{\text{BH}}\right) = w_1^2 \sigma_1^4 \left(1+\rho_1^2\right) + 2w_1^2 \sigma_1^2 \left(1+\rho_1\right)\left(1+\mu_1\right)^2 \qquad (11.54)$$

The serial correlation appears in Equations 11.53 and 11.54 as both a linear term and a square term. The square term is positive but in all likelihood negligibly small.

- Considering the linear term, we note that both variances increase when ρ_1 is positive, i.e. returns are trending, and both variances decrease when ρ_1 is negative, i.e. returns are mean-reverting. In other words, trending returns increase wealth variance, while mean-reverting returns decrease wealth variance.

- Furthermore, the magnitude of change is smaller for the FW portfolio than it is for the BH portfolio, since $\mu_p < \mu_1$. For instance, when ρ_1 is negative, both variances will decline from the levels when ρ_1 is zero. But the variance of the BH portfolio will decline more.

However, the variance of the FW portfolio will always be lower than that of the BH portfolio, no matter what the serial correlation is. This can be seen by considering their difference:

$$\text{var}\left(W_{\text{FW}}\right) - \text{var}\left(W_{\text{BH}}\right) = \left(w_1^2 - 1\right) w_1^2 \sigma_1^4 \left(1+\rho_1^2\right)$$
$$+ 2w_1^2 \sigma_1^2 \left(1+\rho_1\right)\left[\left(1+w_1\mu_1+w_0\mu_0\right)^2 - \left(1+\mu_1\right)^2\right] \qquad (11.55)$$

Both terms in Equation 11.55 are negative when $0 < w_1 < 1$ and $\mu_1 > \mu_0$.

11.7 TERMINAL WEALTH OF LONG-SHORT PORTFOLIOS

Even though the results regarding the expected terminal wealth and the variance are general for any portfolio setting, we have so far directed our analyses mostly to long-only portfolios.

In this section, we specifically examine the case of long-short portfolios, for the effects of portfolio rebalancing on long-short portfolios are markedly different. We have noticed this already from Equation 11.14: for long-short portfolios with two assets and two time periods, the expected terminal wealth of the FW portfolio is actually higher than that of the BH portfolio, regardless of the expected returns of the two assets. This result is completely opposite to the result for long-only portfolios.

It is helpful to understand how this occurs in this simple case. Let us suppose that the portfolio is long Asset 1 and short Asset 2, i.e. $w_1 > 0$ and $w_2 < 0$. Since $w_1 + w_2 = 1$, we know that $w_1 > 1$. Then, at the end of the first period, the average weights are given by Equation 11.13. We write the ratios of new weights to the original weights:

$$\frac{w_1^*}{w_1} = \frac{1 + \mu_1}{1 + \mu_1 - w_2(\mu_1 - \mu_2)}; \frac{w_2^*}{w_2} = \frac{1 + \mu_2}{1 + \mu_2 + w_1(\mu_1 - \mu_2)} \tag{11.56}$$

These ratios measure a portfolio's leverage, and depending on the expected returns of the two assets, leverage can change in different ways. When $\mu_1 > \mu_2$, then the denominators of both ratios in Equation 11.56 will be greater than the numerators, since $w_1 > 1$ and $w_2 < 0$. We then have $1 < w_1^* < w_1$ and $0 > w_2^* > w_2$. Hence, when the portfolio is positioned correctly, i.e. long the asset with a higher expected return and short the asset with a lower expected return, the magnitudes of new portfolio weights will be smaller, resulting in a reduction in portfolio leverage. This is obviously not a desirable outcome in terms of leverage. On the other hand, when $\mu_1 < \mu_2$, the denominators of both ratios will be smaller than the numerators. Hence, when the portfolio is positioned incorrectly, i.e. long the asset with a lower expected return and short the asset with a higher expected return, the magnitudes of new portfolio weights will be greater, resulting in an increase in portfolio leverage, an even worse outcome. Therefore, a BH approach is truly incompatible with a long-short portfolio. Portfolio rebalancing in the FW portfolio, however, would either restore leverage when the portfolio is positioned correctly or cut leverage, in other words stop loss, if the portfolio is positioned incorrectly. This is why the FW approach would yield higher expected terminal wealth than the BH approach.

This result is true for general cases with multiple assets and multiple time periods. However, the proof is not quite as straightforward as in the long-only case.

We consider the expected terminal wealth of a long-short portfolio in which there is only one asset being short, denoted by subscript 1. The assets from 2 to M are long. In other words, $w_1 < 0$, and $w_i > 0, i = 2,\ldots,M$. Then, the expected terminal wealth of the FW portfolio is given by

$$E(W_{FW}) = \left(1 + \sum_{i=2}^{M} w_i \mu_i + w_1 \mu_1\right)^N \tag{11.57}$$

Note that we have separated the long assets and the short asset. The expected terminal wealth of the BH portfolio is

$$E(W_{BH}) = \sum_{i=2}^{M} w_i (1+\mu_i)^N + w_1 (1+\mu_1)^N \qquad (11.58)$$

To prove that $E(W_{FW}) \geq E(W_{BH})$, it suffices to show

$$\left(1 + \sum_{i=2}^{M} w_i \mu_i + w_1 \mu_1\right)^N - w_1 (1+\mu_1)^N \geq \sum_{i=2}^{M} w_i (1+\mu_i)^N \qquad (11.59)$$

To proceed, we assume that the expected returns of all long assets exceed or are equal to the expected return of the short asset, i.e. $\mu_i \geq \mu_1, i = 2,\ldots,M$. Furthermore, $\mu_i > \mu_1$ for some i, since when they are all equal, the expected terminal wealth is the same.

We now write the expected return μ_p as the sum of μ_1 and the excess returns of long assets versus μ_1, i.e.

$$\mu_p = \sum_{i=2}^{M} w_i \mu_i + w_1 \mu_1 = \sum_{i=2}^{M} w_i (\mu_i - \mu_1) + \mu_1 \qquad (11.60)$$

Define

$$\theta_i = \frac{\mu_i - \mu_1}{\sum_{i=2}^{M} w_i (\mu_i - \mu_1)}, i = 2,\ldots,M \qquad (11.61)$$

Then, we have $0 \leq \theta_i \leq 1$ and $\sum_{i=2}^{M} w_i \theta_i = 1$.

In addition, we have

$$\sum_{i=2}^{M} w_i (1-\theta_i) = \sum_{i=2}^{M} w_i - \sum_{i=2}^{M} w_i \theta_i = \sum_{i=2}^{M} w_i - 1 = -w_1 \qquad (11.62)$$

Using these two identities with summations in the left-hand side of Equation 11.59, we have

$$\left(1 + \sum_{i=2}^{M} w_i \mu_i + w_1 \mu_1\right)^N - w_1 (1+\mu_1)^N = \sum_{i=2}^{M} w_i \left[\theta_i (1+\mu_p)^N + (1-\theta_i)(1+\mu_1)^N\right] \qquad (11.63)$$

It can be recognized that the terms in the brackets are a weighted sum of the convex function $f(x) = (1+x)^N$. Using Jensen's inequality, we have

$$\left(1+\sum_{i=2}^{M} w_i \mu_i + w_1 \mu_1\right)^N - w_1(1+\mu_1)^N \geq \sum_{i=2}^{M} w_i \left[1 + \theta_i \mu_p + (1-\theta_i)\mu_1\right]^N \qquad (11.64)$$

Lastly, since

$$\theta_i \mu_p + (1-\theta_i)\mu_1 = \frac{\mu_i - \mu_1}{\sum_{i=2}^{M} w_i(\mu_i - \mu_1)}\left[\sum_{i=2}^{M} w_i(\mu_i - \mu_1) + \mu_1\right]$$

$$+ \left[1 - \frac{\mu_i - \mu_1}{\sum_{i=2}^{M} w_i(\mu_i - \mu_1)}\right]\mu_1 \qquad (11.65)$$

$$= \mu_i - \mu_1 + \frac{\mu_1(\mu_i - \mu_1)}{\sum_{i=2}^{M} w_i(\mu_i - \mu_1)} + \mu_1 - \frac{\mu_1(\mu_i - \mu_1)}{\sum_{i=2}^{M} w_i(\mu_i - \mu_1)} = \mu_i$$

we arrive at Equation 11.59, which proves that $E(W_{FW}) \geq E(W_{BH})$ for this long-short portfolio. We make several remarks about the proof.

- First, note that the proof is also applicable if the expected returns of all long assets are less than or equal to the expected return of the short asset.

- Second, the inequality does not seem to hold in general for cases in which some assets' expected returns are higher than that of the short asset, whereas other assets' expected returns are lower than that of the short asset.

The proof can be extended to cases in which there are multiple assets rather than one asset that are short. Without loss of generality, we assume that the first M_S assets are shorts, i.e. $w_i < 0, i = 1,\dots, M_S$. The rest of the assets are longs, with $w_j > 0, j = M_S + 1,\dots, M$. We use index i for short assets and index j for long assets. To prove that $E(W_{FW}) \geq E(W_{BH})$, it suffices to show that

$$\left(1+\sum_{i=1}^{M_s} w_i \mu_i + \sum_{j=M_s+1}^{M} w_j \mu_j\right)^N - \sum_{i=1}^{M_s} w_i(1+\mu_i)^N \geq \sum_{j=M_s+1}^{M} w_j(1+\mu_j)^N \qquad (11.66)$$

We define the weighted expected return of shorted assets:

$$\mu_s = \frac{\sum_{i=1}^{M_s} w_i \mu_i}{W_s}, \text{ with } W_s = \sum_{i=1}^{M_s} w_i \quad (11.67)$$

The weight W_s is the total short position of the portfolio. Then, the expected portfolio return can be rewritten as

$$\mu_p = \left(\sum_{i=1}^{M_s} w_i\right)\mu_s + \sum_{j=M_s+1}^{M} w_j\left(\mu_j - \mu_s\right) + \left(\sum_{j=M_s+1}^{M} w_j\right)\mu_s = \sum_{j=M_s+1}^{M} w_j\left(\mu_j - \mu_s\right) + \mu_s \quad (11.68)$$

We now assume that the expected return of all long assets is higher than or equal to the average return of shorts; that is, $\mu_j \geq \mu_s$, $j = M_s + 1, \ldots, M$. We first consider the case in which $\mu_j > \mu_s$ for some j. Define

$$\theta_j = \frac{\mu_j - \mu_s}{\sum_{i=M_s+1}^{M} w_j\left(\mu_j - \mu_s\right)}, j = M_s + 1, \ldots, M \quad (11.69)$$

Then, $0 \leq \theta_j \leq 1$ and $\sum_{j=M_s+1}^{M} w_j\theta_j = 1$

In addition, we have

$$\sum_{j=M_s+1}^{M} w_j\left(1 - \theta_j\right) = \sum_{j=M_s+1}^{M} w_j - \sum_{j=M_s+1}^{M} w_j\theta_j = \sum_{j=M_s+1}^{M} w_j - 1 = -\sum_{i=1}^{M_s} w_i = -W_s \quad (11.70)$$

Or alternatively,

$$\frac{\sum_{j=M_s+1}^{M} w_j\left(1 - \theta_j\right)}{W_s} = -1 \quad (11.71)$$

The left-hand side of Equation 11.66 can be rewritten as

$$\left(\sum_{j=M_s+1}^{M} w_j\theta_j\right)\left(1 + \mu_p\right)^N + \left[\frac{\sum_{j=M_s+1}^{M} w_j\left(1 - \theta_j\right)}{W_s}\right]\sum_{i=1}^{M_s} w_i\left(1 + \mu_i\right)^N \quad (11.72)$$

$$= \sum_{j=M_s+1}^{M} w_j\left\{\theta_j\left(1 + \mu_p\right)^N + \left(1 - \theta_j\right)\sum_{i=1}^{M_s} \frac{w_i}{W_s}\left(1 + \mu_i\right)^N\right\}$$

For every j, the terms inside the brackets are weighted sums of the convex function $f(x) = (1+x)^N$ at $M_s + 1$ points, since their coefficients are all non-negative and they sum up to unity; that is,

$$\theta_j + (1-\theta_j) \sum_{i=1}^{M_s} \frac{w_i}{W_s} = \theta_j + (1-\theta_j) = 1 \qquad (11.73)$$

Then by Jensen's inequality, we have

$$\sum_{j=M_s+1}^{M} w_j \left\{ \theta_j (1+\mu_p)^N + (1-\theta_j) \sum_{i=1}^{M_s} \frac{w_i}{W_s} (1+\mu_i)^N \right\} \geq$$

$$\qquad (11.74)$$

$$\sum_{j=M_s+1}^{M} w_j \left[1 + \theta_j \mu_p + (1-\theta_j) \sum_{i=1}^{M_s} \frac{w_i}{W_s} \mu_i \right]^N = \sum_{j=M_s+1}^{M} w_j (1+\mu_j)^N$$

The last step of equality is based on the fact that the argument of the power function is

$$\theta_j \mu_p + (1-\theta_j) \sum_{i=1}^{M_s} \frac{w_i}{W_s} \mu_i$$

$$= \frac{\mu_j - \mu_s}{\sum_{i=M_s+1}^{M} w_j (\mu_j - \mu_s)} \left[\sum_{j=M_s+1}^{M} w_j (\mu_j - \mu_s) + \mu_s \right]$$

$$+ \left[1 - \frac{\mu_j - \mu_s}{\sum_{i=M_s+1}^{M} w_j (\mu_j - \mu_s)} \right] \mu_s \qquad (11.75)$$

$$= \mu_j - \mu_s + \frac{(\mu_j - \mu_s)\mu_s}{\sum_{i=M_s+1}^{M} w_j (\mu_j - \mu_s)} + \mu_s - \frac{(\mu_j - \mu_s)\mu_s}{\sum_{i=M_s+1}^{M} w_j (\mu_j - \mu_s)} = \mu_j$$

This completes the proof. We now consider the special case in which $\mu_j = \mu_s$, $j = M_s + 1, \ldots, M$. By Equation 11.67, we have $\mu_p = \mu_s$. And Equation 11.66 reduces to

$$(1+\mu_s)^N - \sum_{i=1}^{M_s} w_i (1+\mu_i)^N \geq (1+\mu_s)^N \sum_{j=M_s+1}^{M} w_j \qquad (11.76)$$

Rearranging the terms yields

$$\sum_{i=1}^{M_s} \frac{w_i}{W_s}\left(1+\mu_i\right)^N \geq \left(1+\mu_s\right)^N \tag{11.77}$$

This inequality is true by Jensen's inequality and the definition of μ_s.

The picture on the variance of terminal wealth for long-short portfolios is less clear. In the case of a long-only portfolio, we have shown that in many cases, FW portfolios have lower variances of terminal wealth than BH portfolios. For long-short portfolios, should we then expect a higher variance of terminal wealth, since we have seen that the opposite is true for the expected value of terminal wealth? The answer is not necessarily. This is because for long-short portfolios, there are two competing factors in determining the variance of FW portfolios versus that of BH portfolios. Some long-short portfolios can be thought of as levered "long-only" portfolios. In such cases, portfolio rebalancing can be thought of as portfolio rebalancing of the "long-only" portfolio plus leverage rebalancing. The first factor is thus portfolio leverage. As we have argued earlier, portfolio rebalancing keeps FW portfolios' leverage constant, whereas BH portfolios' leverage will change over time. When portfolios' long and short positions are correctly aligned with expected returns, the leverage will decline over time for BH portfolios. Declining leverage leads to a lower variance of terminal wealth. Hence, under these conditions, portfolio leverage will result in a higher variance of terminal wealth for the FW portfolios, as it results in higher expected terminal wealth. The second factor is the inherent advantage of portfolio diversification for the FW portfolios that we have demonstrated in long-only portfolios. When rebalancing long-short portfolios, we still keep the benefit of a lower variance for the "long-only" portion of the FW portfolios. The relative magnitude of portfolio leverage and the benefit of long-only portfolio rebalancing will determine whether the variance of FW long-short portfolios is lower or higher than that of BH long-short portfolios.

Using these insights, we revisit some of the previous long-only examples in the long-short setting. First, consider the two-asset example in which one asset is risky and the other is risk-free. In a long-short setting, the weight of the risky asset would be greater than 1, i.e. $w_1 > 1$, and the weight of the risk-free asset would be less than 0, i.e. $w_0 < 0$. In this case, there is only one risky asset, so only the leverage effect is present. We have a complete role reversal compared with the long-only case—the FW portfolio has both a higher expected terminal wealth and a higher variance of terminal wealth. This is mathematically evident from Equations 11.53 and 11.54. When $w_1 > 1$ and $w_1\mu_1 + w_0\mu_0 > \mu_1$, we have $\text{var}(W_{\text{FW}}) > \text{var}(W_{\text{BH}})$.

For portfolios with more than one risky asset, the situation is more complicated and beyond the scope of the book.

APPENDIX 11.A RISK-ADJUSTED WEALTH OF TWO-ASSET LONG-ONLY PORTFOLIOS

We prove that for the two-asset case, the risk-adjusted terminal is higher for the FW portfolio than for the BH portfolio. We start with the reciprocal of this ratio squared. For the FW portfolio, we have

$$\left[\frac{\text{std}(W_{\text{FW}})}{E(W_{\text{FW}})}\right]^2 = \sum_{n=1}^{N}\left[C_N^n\left(\frac{w_1\sigma_1}{1+w_1\mu_1+w_0\mu_0}\right)^{2n}\right] \tag{11.78}$$

For the BH portfolio, we have

$$\left[\frac{\text{std}(W_{\text{BH}})}{E(W_{\text{BH}})}\right]^2 = \frac{w_1^2(1+\mu_1)^{2N}}{\left[w_1(1+\mu_1)^N + w_0(1+\mu_0)^N\right]^2}\sum_{n=1}^{N}\left[C_N^n\left(\frac{\sigma_1}{1+\mu_1}\right)^{2n}\right] \tag{11.79}$$

For the FW portfolio, we rewrite Equation 11.78 as

$$\left[\frac{\text{std}(W_{\text{FW}})}{E(W_{\text{FW}})}\right]^2 = \sum_{n=1}^{N}\left[C_N^n\left(\frac{\sigma_1}{1+\mu_1}\right)^{2n}\left(1+\frac{w_0}{w_1}\frac{1+\mu_0}{1+\mu_1}\right)^{-2n}\right] \tag{11.80}$$

Let $y = (1+\mu_0)/(1+\mu_1)$. Then, $0 < y < 1$, and Equation 11.80 becomes

$$\left[\frac{\text{std}(W_{\text{FW}})}{E(W_{\text{FW}})}\right]^2 = \sum_{n=1}^{N}\left[C_N^n\left(\frac{\sigma_1}{1+\mu_1}\right)^{2n}\left(1+\frac{w_0}{w_1}y\right)^{-2n}\right] \tag{11.81}$$

For the BH portfolio, the ratio squared is

$$\left[\frac{\text{std}(W_{\text{BH}})}{E(W_{\text{BH}})}\right]^2 = \left(1+\frac{w_0}{w_1}y^N\right)^{-2}\sum_{n=1}^{N}\left[C_N^n\left(\frac{\sigma_1}{1+\mu_1}\right)^{2n}\right] \tag{11.82}$$

Since $1 + (w_0 / w_1) y > 1$ and $0 < y < 1$, we have

$$\left(1 + \frac{w_0}{w_1} y\right)^n \geq 1 + \frac{w_0}{w_1} y > 1 + \frac{w_0}{w_1} y^N \tag{11.83}$$

Therefore,

$$\left(1 + \frac{w_0}{w_1} y\right)^{-2n} < \left(1 + \frac{w_0}{w_1} y^N\right)^{-2} \quad \text{for } n = 1, \dots, N \tag{11.84}$$

Using this inequality in Equations 11.81 and 11.82, we obtain

$$\left[\frac{\text{std}(W_{\text{FW}})}{E(W_{\text{FW}})}\right]^2 < \left[\frac{\text{std}(W_{\text{BH}})}{E(W_{\text{BH}})}\right]^2 \tag{11.85}$$

This proves Equation 11.42.

APPENDIX 11.B EXPECTED TERMINAL WEALTH AND VARIANCE WITH SERIAL CORRELATIONS

We derive analytic results of the expected terminal wealth and variance of terminal wealth when there are serial correlations between asset returns from different time periods. To keep the problem analytically tractable, we only consider the case with two time periods, i.e. $N = 2$. While not a general case, a horizon with two time periods still presents an opportunity for portfolio rebalancing for the FW approach and therefore, an opportunity to compare the FW approach and the BH approach.

We keep the number of assets general at M. Then, the terminal wealth of the FW portfolio is

$$W_{\text{FW}} = \left(1 + \sum_{i=1}^{M} w_i r_{i1}\right)\left(1 + \sum_{i=1}^{M} w_i r_{i2}\right) \tag{11.86}$$

and the terminal wealth of the BH portfolio is

$$W_{\text{BH}} = w_1 (1 + r_{11})(1 + r_{12}) + \cdots + w_M (1 + r_{M1})(1 + r_{M2})$$

$$= \sum_{i=1}^{M} w_i (1 + r_{i1})(1 + r_{i2}) \tag{11.87}$$

We denote the covariance between return of asset i in Period 1 and return of asset j in Period 2 as $\sigma_{ij}^{1,2}$, with superscript 1 and 2 for the time periods.

$$E\left[\left(r_{i1}-\mu_i\right)\left(r_{j2}-\mu_j\right)\right]=\sigma_{ij}^{1,2} \text{ for } i,j=1,\ldots,M \tag{11.88}$$

These serial covariances can be summarized in a matrix:

$$\Sigma^{1,2}=\begin{pmatrix} \sigma_{11}^{1,2} & \cdots & \sigma_{1M}^{1,2} \\ \vdots & \ddots & \vdots \\ \sigma_{M1}^{1,2} & \cdots & \sigma_{MM}^{1,2} \end{pmatrix} \tag{11.89}$$

In contrast to the covariance matrix for returns within the same time period, $\Sigma^{1,2}$ is not symmetric, since $\sigma_{ij}^{1,2}$ and $\sigma_{ji}^{1,2}$ have different lead-lag relations when i and j are different.

Both the expectation and the variance of terminal wealth will depend on these serial correlations. We consider the expectation first. For the FW portfolio, we have

$$E\left(W_{\text{FW}}\right)=E\left[\left(1+\sum_{i=1}^{M}w_i r_{i1}\right)\left(1+\sum_{i=1}^{M}w_i r_{i2}\right)\right] \tag{11.90}$$

The expectation of a product of two random variables equals the product of their means plus the covariance, i.e. $E\left(xy\right)=E\left(x\right)E\left(y\right)+\text{cov}\left(x,y\right)$. Hence,

$$E\left(W_{\text{FW}}\right)=E\left(1+\sum_{i=1}^{M}w_i r_{i1}\right)E\left(1+\sum_{i=1}^{M}w_i r_{i1}\right)+\text{cov}\left(1+\sum_{i=1}^{M}w_i r_{i1},1+\sum_{i=1}^{M}w_i r_{i2}\right) \tag{11.91}$$

The first term involves expected returns of the FW portfolio, and the second term results in covariances between returns in Periods 1 and 2. We obtain

$$E\left(W_{\text{FW}}\right)=\left(1+\sum_{i=1}^{M}w_i\mu_i\right)^2+\sum_{i,j=1}^{M}w_i w_j\sigma_{ij}^{1,2}=\left(1+\mu_p\right)^2+\vec{w}'\Sigma^{1,2}\vec{w} \tag{11.92}$$

It is easy to see that when there is no serial correlation, Equation 11.92 reduces to the previous result for the case $N = 2$. The second term reflects the effect of serial correlation on the terminal wealth. We note that when the weights are positive, the negative serial correlations lead to lower expected terminal wealth, whereas positive serial correlations lead to higher expected terminal wealth.

For the BH portfolio, the expected terminal wealth is given by

$$E\left(W_{BH}\right) = \sum_{i=1}^{M} w_i E\left[\left(1+r_{i1}\right)\left(1+r_{i2}\right)\right] \tag{11.93}$$

Since $E\left[\left(1+r_{i1}\right)\left(1+r_{i2}\right)\right] = \left(1+\mu_i\right)^2 + \sigma_{ii}^{1,2}$, the expectation of terminal wealth of the BH portfolio is

$$E\left(W_{BH}\right) = \sum_{i=1}^{M} w_i \left(1+\mu_i\right)^2 + \sum_{i=1}^{M} w_i \sigma_{ii}^{1,2} \tag{11.94}$$

We note that the first term is the expectation absent any serial correlation. The second term, which reflects the effect of serial correlations, only involves auto serial correlations of individual assets. Also, another notable difference between Equations 11.94 and 11.92 is that in Equation 11.94, the serial correlations are multiplied by weights, while in Equation 11.92, they are multiplied by weights "squared."

We next consider the variance of terminal wealth, for which we analyze the expectation of terminal wealth squared. For the FW portfolio, we have

$$W_{FW}^2 = \left(1+\sum_{i=1}^{M} w_i r_{i1}\right)^2 \left(1+\sum_{i=1}^{M} w_i r_{i2}\right)^2 \triangleq \left(1+r_{p1}\right)^2 \left(1+r_{p2}\right)^2 \tag{11.95}$$

We have defined portfolio return r_{p1} and r_{p2} in Periods 1 and 2, respectively. Taking expectation of the equation yields

$$E\left(W_{FW}^2\right) = E\left[\left(1+r_{p1}\right)^2\right] E\left[\left(1+r_{p2}\right)^2\right] + \text{cov}\left[\left(1+r_{p1}\right)^2, \left(1+r_{p2}\right)^2\right] \tag{11.96}$$

The two expectations in the first terms are the same, and the results are

$$E\left[\left(1+r_{p1}\right)^2\right] = E\left[\left(1+r_{p2}\right)^2\right] = \left(1+\mu_p\right)^2 + \sigma_p^2 \tag{11.97}$$

To analyze the second covariance term in Equation 11.96, we now assume that asset returns are normally distributed, so we can use the following statistical identity. Suppose both x and y are normally distributed random variables; then

$$\text{cov}\left(x^2, y^2\right) = 4\text{E}(x)\text{E}(y)\text{cov}(x, y) + 2\left[\text{cov}(x, y)\right]^2 \quad (11.98)$$

Based on this identity, we obtain

$$\text{cov}\left[\left(1 + r_{p1}\right)^2, \left(1 + r_{p2}\right)^2\right] = 4\left(1 + \mu_p\right)^2 \text{cov}\left(r_{p1}, r_{p2}\right) + \left[\text{cov}\left(r_{p1}, r_{p2}\right)\right]^2$$

$$= 4\left(1 + \mu_p\right)^2 \left(\vec{w}' \Sigma^{1,2} \vec{w}\right) + \left[\left(\vec{w}' \Sigma^{1,2} \vec{w}\right)\right]^2 \quad (11.99)$$

Substituting Equations 11.99 and 11.97 into Equation 11.96 yields

$$\text{E}\left(W_{FW}^2\right) = \left[\left(1 + \mu_p\right)^2 + \sigma_p^2\right]^2 + 4\left(1 + \mu_p\right)^2 \left(\vec{w}' \Sigma^{1,2} \vec{w}\right) + \left[\left(\vec{w}' \Sigma^{1,2} \vec{w}\right)\right]^2 \quad (11.100)$$

To obtain the variance of terminal wealth, we subtract the square of Equation 11.92 from Equation 11.100 to obtain

$$\text{var}\left(W_{FW}\right) = \sigma_p^4 + 2\sigma_p^2 \left(1 + \mu_p\right)^2 + \left[\left(\vec{w}' \Sigma^{1,2} \vec{w}\right)\right]^2 + 2\left(1 + \mu_p\right)^2 \left(\vec{w}' \Sigma^{1,2} \vec{w}\right) \quad (11.101)$$

Notice that the first two terms match the variance in the case of no serial correlation for $N = 2$, and the last two terms are the effects of serial correlations. In addition, we note that since $\sigma_p^2 = \vec{w}' \Sigma \vec{w}$, Equation 11.101 is completely symmetric with respect to covariances between asset returns in the same time period and covariances between asset returns in the different time periods.

We now proceed to find the expectation of terminal wealth of the BH portfolio squared. Since

$$W_{BH}^2 = \left[\sum_{i=1}^{M} w_i \left(1 + r_{i1}\right)\left(1 + r_{i2}\right)\right]^2 = \sum_{i,j=1}^{M} w_i w_j \left(1 + r_{i1}\right)\left(1 + r_{i2}\right)\left(1 + r_{j1}\right)\left(1 + r_{j2}\right) \quad (11.102)$$

we have

$$\text{E}\left(W_{BH}^2\right) = \sum_{i,j=1}^{M} w_i w_j \text{E}\left[\left(1 + r_{i1}\right)\left(1 + r_{j1}\right)\left(1 + r_{i2}\right)\left(1 + r_{j2}\right)\right] \quad (11.103)$$

To evaluate the expectation of a product of four factors, we simplify the notation with $x_1 = 1 + r_{i1}$, $y_1 = 1 + r_{j1}$, $x_2 = 1 + r_{i2}$, $y_2 = 1 + r_{j2}$ and write it as

$$E\left[(1+r_{i1})(1+r_{j1})(1+r_{i2})(1+r_{j2})\right] \triangleq E(x_1 y_1 x_2 y_2) \qquad (11.104)$$

A first step is to recognize that $E(x_1 y_1 x_2 y_2) = E(x_1 y_1) E(x_2 y_2) + \text{cov}(x_1 y_1, x_2 y_2)$. The first term consists of expectation of returns in the same time period, which has been analyzed before. We have

$$E(x_1 y_1) = E(x_2 y_2) = (1+\mu_i)(1+\mu_j) + \sigma_{ij} \qquad (11.105)$$

The second covariance term between the returns of different time periods comes from the serial covariances. We shall use the following identity for its evaluation:

$$\text{cov}(x_1 y_1, x_2 y_2) = \text{cov}(x_1, x_2)\text{cov}(y_1, y_2) + \text{cov}(x_1, y_2)\text{cov}(y_1, x_2)$$
$$+E(x_1)E(x_2)\text{cov}(y_1, y_2) + E(x_1)E(y_2)\text{cov}(y_1, x_2) \qquad (11.106)$$
$$+E(y_1)E(x_2)\text{cov}(x_1, y_2) + E(y_1)E(y_2)\text{cov}(x_1, x_2)$$

Substituting x's and y's, we have

$$\text{cov}(x_1 y_1, x_2 y_2) = \sigma_{ii}^{1,2}\sigma_{jj}^{1,2} + \sigma_{ij}^{1,2}\sigma_{ji}^{1,2} + (1+\mu_i)^2 \sigma_{jj}^{1,2} + (1+\mu_i)(1+\mu_j)\sigma_{ij}^{1,2}$$
$$+(1+\mu_j)(1+\mu_i)\sigma_{ji}^{1,2} + (1+\mu_j)^2 \sigma_{ii}^{1,2} \qquad (11.107)$$

Hence,

$$E\left[(1+r_{i1})(1+r_{j1})(1+r_{i2})(1+r_{j2})\right] = \left[(1+\mu_i)(1+\mu_j) + \sigma_{ij}\right]^2$$
$$+ \sigma_{ii}^{1,2}\sigma_{jj}^{1,2} + \sigma_{ij}^{1,2}\sigma_{ji}^{1,2}$$
$$+(1+\mu_i)^2 \sigma_{jj}^{1,2} + (1+\mu_i)(1+\mu_j)\sigma_{ij}^{1,2} \qquad (11.108)$$
$$+(1+\mu_j)(1+\mu_i)\sigma_{ji}^{1,2} + (1+\mu_j)^2 \sigma_{ii}^{1,2}$$

Using this result in Equation 11.103 gives

$$
E\left(W_{BH}^2\right) = \sum_{i,j=1}^{M} w_i w_j \left[\left(1+\mu_i\right)\left(1+\mu_j\right)+\sigma_{ij}\right]^2 + \left(\sum_{i=1}^{M} w_i \sigma_{ii}^{1,2}\right)^2 + \sum_{i,j=1}^{M} w_i w_j \sigma_{ij}^{1,2} \sigma_{ji}^{1,2}
$$

$$
+2\left[\sum_{i=1}^{M} w_i \left(1+\mu_i\right)^2\right]\left[\sum_{i=1}^{M} w_i \sigma_{ii}^{1,2}\right] + 2\sum_{i,j=1}^{M} w_i w_j \left(1+\mu_j\right)\left(1+\mu_i\right)\sigma_{ij}^{1,2} \tag{11.109}
$$

From Equation 11.94, we have

$$
\left[E\left(W_{BH}\right)\right]^2 = \left[\sum_{i=1}^{M} w_i \left(1+\mu_i\right)^2\right]^2 + 2\left[\sum_{i=1}^{M} w_i \left(1+\mu_i\right)^2\right]\left[\sum_{i=1}^{M} w_i \sigma_{ii}^{1,2}\right]
$$

$$
+\left(\sum_{i=1}^{M} w_i \sigma_{ii}^{1,2}\right)^2 \tag{11.110}
$$

Subtracting this expression from Equation 11.109 yields

$$
\text{var}\left(W_{BH}\right) = \sum_{i,j=1}^{M} w_i w_j \left[2\left(1+\mu_i\right)\left(1+\mu_j\right)\sigma_{ij}+\sigma_{ij}^2\right]
$$

$$
+\sum_{i,j=1}^{M} w_i w_j \left[2\left(1+\mu_i\right)\left(1+\mu_j\right)\sigma_{ij}^{1,2}+\sigma_{ij}^{1,2}\sigma_{ji}^{1,2}\right] \tag{11.111}
$$

This is the result of the variance of terminal wealth for the BH portfolio, which consists of two summations, the first over return covariances within the same time period and the second over return covariances between two different time periods. Again, the two summations are symmetric.

PROBLEMS

11.1. Derive Equation 11.14.

11.2. Derive Equation 11.51.

11.3. Prove Equation 11.60.

CHAPTER 12

Threshold Rebalancing

T HE STUDY OF PORTFOLIO rebalancing so far has made the implicit assumption that rebalancing occurs at a regular interval. In the example of sector portfolios, rebalancing is done on an annual basis. The use of a subscript for time also implies regularity, even though this need not be the case. In some parts of the text, we have also used a normal distribution to describe return samples from different periods, with an implicit assumption that all periods are of the same length. In practice, regular portfolio rebalancing is often carried out on a calendar basis.

Portfolio rebalancing need not be on a regular basis. Another kind of portfolio rebalancing is threshold rebalancing, which only occurs if changes in portfolio weights cross a prescribed threshold. For example, suppose the original portfolio weights are 50/50 for a portfolio with two assets. We can let the portfolio drift as long as the weights do not deviate by more than 5%. The 5% threshold means that we have a buy-and-hold (BH) portfolio as long as the portfolio weights are within 45/55 and 55/45. However, whenever the weights are outside the threshold, we rebalance the portfolio to the original 50/50 portfolio.

Compared with regular rebalancing, threshold rebalancing might bring several potential benefits. First, it could potentially reduce turnover and transaction costs. It is possible that small deviations from the original portfolio weights are pure noise, and rebalancing small portfolio drifts has no benefits. From a different perspective, when the deviation of portfolios is small, the tracking error between the fixed-weight (FW) and BH portfolios is small. Hence, threshold rebalancing simply forfeits the opportunity to take action against the BH portfolio. Second, it is conceivable that the probability of return mean reversion will increase when return differences are among assets, which could push the weights across the threshold. If we only rebalance then, the rebalancing alpha might be greater than with a regular portfolio rebalancing. Another characteristic of threshold rebalancing is that the portfolios will be monitored more frequently to check when the threshold is met. The period is often much shorter than with calendar rebalancing.

12.1 RETURN DISPERSION OR WEIGHT DISPERSION AS A THRESHOLD

While it is rather intuitive to use changes in portfolio weights as thresholds for portfolios of a few assets, this is less practical for portfolios of multiple securities. For instance, suppose we set 5% in deviation of portfolio weights as the rebalancing threshold for a portfolio of four assets. Then, deviations of 4%, 4%, −4%, and −4% would not trigger a rebalance. However, collectively, these deviations represent substantial drifts from the original weights. A threshold of 5% is a single point threshold. A threshold based on statistics would be a better choice.

A natural choice is the dispersion, or cross-sectional standard deviation of weight shifts. According to Equation 12.1, the dispersion of weight shifts is related to the dispersion of asset returns. When it is an equally weighted portfolio, we have

$$\text{std}\left(\triangle \mathbf{w}\right) = \frac{\text{std}\left(\mathbf{r}\right)}{M\left(1+R\right)} \tag{12.1}$$

When the portfolio is not equally weighted, the dispersion of weight shifts, and the dispersion of returns is not related explicitly. In this chapter, we shall only study threshold rebalancing using the dispersion of asset returns for equally weighted S&P 500 sector portfolios, as discussed previously. It is simple to extend the methodology to threshold rebalancing with dispersion of weight changes.

Does cross-sectional dispersion of sector returns have any impact on subsequent behavior of sector returns? Specifically, do higher dispersions lead to a higher likelihood of mean reversion, which would lend support to portfolio rebalancing? We examine this relationship using monthly returns of sector portfolios studied in earlier chapters. As mentioned before, the period used to implement threshold rebalancing

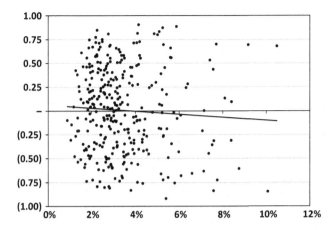

FIGURE 12.1 Scatter plot of monthly return dispersion (October 1989 to December 2016) of serial cross-sectional correlation of lag one month.

is often shorter than for calendar rebalancing. In practice, it could be weekly, daily, or even by minutes. Here, we use the monthly frequency for illustration.

Figure 12.1 displays the scatter plot of cross-sectional dispersion of sector return in a given month versus the correlation between those sector returns and the sector returns of the subsequent month. The dispersion of monthly returns ranges from 1% to 11%. Overall, the average of this Lag 1 serial cross-correlation is an insignificant 0.005. However, there is a negative correlation between the correlations and the dispersion of returns of the leading month. As the trend line indicates, when the dispersion is small, the cross-sectional correlation is slightly positive on average, implying momentum; when the dispersion is large, the cross-sectional correlation is slightly negative on average, implying mean reversion. While this observation is not a rigorous proof that threshold rebalancing based on the dispersion of returns will be better than monthly rebalancing, since rebalancing alpha is not a sole function of Lag 1 serial cross-sectional correlations of returns, it does provide a motivation to investigate threshold rebalancing.

12.2 NUMERICAL SIMULATION OF THRESHOLD REBALANCING

To test the efficacy of threshold rebalancing using cross-sectional return dispersion, we carry out numerical simulations with different thresholds based on monthly returns of sector portfolios. We make a few remarks about the setting.

- First, we note that the dispersion is not the dispersion of returns in a month. Rather, it is the dispersion of cumulative returns since the last rebalancing. Each rebalance resets the portfolio to the original weights and resets the calculation of cumulative returns.

- Second, at two extremes, both the FW portfolio (monthly rebalance) and the BH portfolio are special cases of threshold rebalancing. For the former, we set the value of threshold to be zero. For the latter, we set the value of threshold to be infinite (or a very large number).

- For a more realistic comparison, we should take into account turnover and resulting transaction costs. A BH portfolio has no turnover, and a FW portfolio would have the most turnover. Portfolios based on threshold rebalancing would lie between the two.

Table 12.1 displays return statistics for these two portfolios from October 1989 to December 2016. The FW portfolio outperformed the BH portfolio net of transaction costs by 38 basis points per annum, and their volatilities are quite similar. As a result, the FW portfolio has a slightly higher Sharpe ratio. However, the FW portfolio is not a good active strategy relative to the BH portfolio, because the tracking error between the two is quite high relative to the rebalancing alpha. Over the entire period, the annualized tracking error is 1.84%, resulting in an information ratio (IR) less than 0.2.

The tracking error has significant variation over time, as indicated by the 12-month rolling tracking error in Figure 12.2. For most of the time, the tracking error is around 1% or lower. However, for two periods, one around the tech bubble of 2000 and the other around the global financial crisis, the tracking error spikes sharply due to both the difference in portfolio weights and large return dispersions among the sectors. It is during the tech bubble that the FW portfolio underperforms the BH portfolio significantly. Figure 12.3 plots the ratio of cumulative returns of the FW to the BH portfolio. The ratio dropped substantially from 1996 to 1999 and snapped back instantly when the bubble finally burst. This episode demonstrates that a long investment horizon is often necessary for portfolio rebalancing to bear fruit.

We now turn to threshold rebalancing with cross-sectional return dispersion. Table 12.2 displays the result of such a case with the threshold value of 5%. Compared with the FW portfolio, which is rebalanced every month, this threshold (TH) portfolio has slightly higher net return and

TABLE 12.1   Return Statistics of FW and BH Sector Portfolios with Monthly Rebalancing Frequency

	FW	BH
Return	9.93%	9.55%
Volatility	13.58%	13.50%
Sharpe	0.49	0.47
Turnover	32%	0%

Note: The Turnover Is Annualized Two-Way Turnover and the Transaction Cost is 50 Basis Points for 100% Turnover.

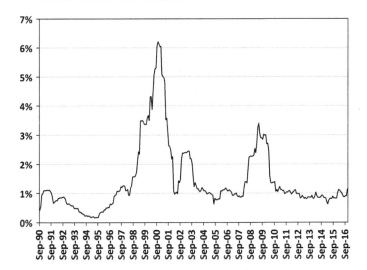

FIGURE 12.2   Rolling 12-month annualized tracking error between FW and BH portfolios.

similar risk with an improvement in Sharpe ratio. We also note that the turnover for the threshold rebalancing is lower at 22% compared with that of the FW portfolio at 32%. Table 12.2 also shows the performance of the TH portfolio relative to the FW portfolio. In the active space, the TH portfolio has an excess return of 8 basis points and tracking error of 13 basis points, resulting in an IR of 0.61.

Figure 12.4 plots the ratio of cumulative returns of the TH to the FW portfolio. The ratio increases steadily over time with one drawdown period from 1998 to 2001. It is interesting to note that relative to the FW portfolio, the TH portfolio is mostly a conditional momentum strategy. In other words, when the return dispersion is less than 5%, the TH portfolio will be a BH portfolio, and when the dispersion is greater than 5%, the TH portfolio is rebalanced back to the FW portfolio. Our results indicate that during

FIGURE 12.3   Ratio of cumulative return of FW portfolio to cumulative return of BH portfolio.

TABLE 12.2   Return Statistics of Threshold Rebalancing with 5% Return Dispersion and Its Relative Performance versus the FW Portfolio

	TH	TH less FW
Return	10.02%	0.08%
Volatility	13.56%	0.13%
Sharpe	0.50	0.61
Turnover	22%	

these BH periods, the TH portfolio outperforms the FW portfolio. Thus, it seems that when dispersion is less than 5%, the momentum strategy (let the winners ride) adds value. From another perspective, the TH portfolio in essence is a truncated BH portfolio. If the threshold is raised high enough, the TH portfolio will become the BH portfolio. Since we know that the BH portfolio underperforms the FW portfolio, this implies that the underperformance must have occurred during the period in which the dispersion was above 5%.

To simulate a wide range of TH portfolios, we let the dispersion threshold vary from 0% (FW) to 50% with an increment of 1%. As the threshold increases, we rebalance less often, and the average turnover declines. Figure 12.5 displays return, volatility, and Sharpe ratio of those 51 portfolios. As the threshold increases, the return increases slightly, while the risk decreases slightly. As a result, the Sharpe ratio increases from 0.49 to 0.54

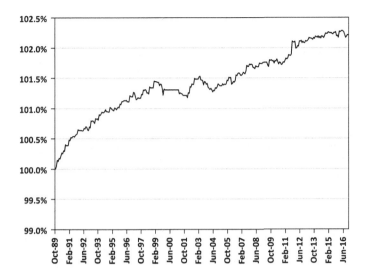

FIGURE 12.4   Ratio of cumulative return of TH portfolio to cumulative return of FW portfolio.

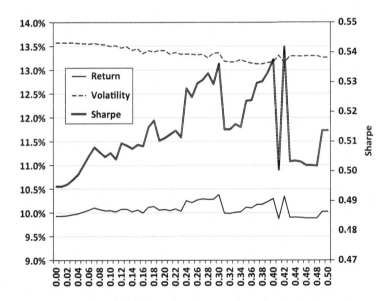

FIGURE 12.5   Return, volatility, and Sharpe ratio of TH portfolios as a function of the threshold.

as the threshold increases from 0% to 30%. However, beyond 30% dispersion, the returns of TH portfolios become unstable, while the risks remain smooth. As a result, the Sharpe ratio is also choppy. The reason is that when the threshold is high, the rebalancing becomes rare and far-between.

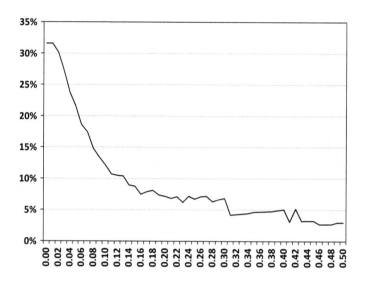

FIGURE 12.6 Average annual turnover of TH portfolios as a function of the threshold.

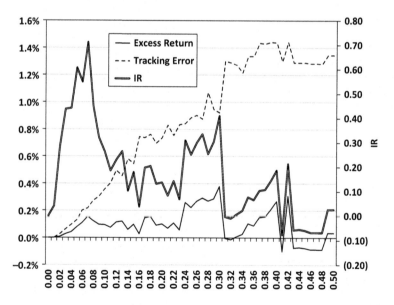

FIGURE 12.7 Excess return, tracking error, and information ratio of TH portfolios against the FW portfolio as a function of the threshold.

A slight change in rebalancing schedule can induce significant change in the underlying TH portfolio, causing the returns and Sharpe ratio to be discontinuous. Figure 12.6 plots the average annual turnover of these TH portfolios. It declines from 30% (FW portfolio) to less than 5% when the threshold goes beyond 30%.

Finally, Figure 12.7 displays excess return, tracking error, and IR of the TH portfolios relative to the FW portfolio. When the threshold is below 30%, the excess return is always positive. However, the tracking error increases monotonically as the TH portfolio is rebalanced less often. Consequently, IR becomes very low. However, it does reach a maximum of 0.7 when the dispersion is 6%, which is quite attractive. In real time, it is probably hard to determine what threshold would yield high IR in practice without hindsight.

Bibliography

Almadi, H., D.E. Rapach, and A. Suri, 2014, "Return predictability and dynamic asset allocation: How often should investors rebalance." *Journal of Portfolio Management*, vol. 40, no. 4: pp 16–27.

Becker, Robert A., "The variance drain and Jensen's inequality" (March 19, 2012). CAEPR Working Paper No. 2012-004. Available at SSRN: https://ssrn.com/abstract=2027471 or http://dx.doi.org/10.2139/ssrn.2027471

Bernstein, W.J. and D. Wilkinson "Diversification, rebalancing, and the geometric mean frontier" (November 24, 1997). Available at SSRN: https://ssrn.com/abstract=53503 or http://dx.doi.org/10.2139/ssrn.53503

Booth, D.G. and E.F. Fama, 1992, "Diversification returns and asset contribution." *Financial Analyst Journal*, vol. 48, no. 3: pp 26–32.

Bouchey, P., V. Nemtchinov, A. Paulsen, and D. Stein, 2012, "Volatility harvesting: Why does diversifying and rebalancing create portfolio growth?" *The Journal of Wealth Management*, vol. 15, no. 2: pp 26–35.

Bouchey, P., V. Nemtchinov, and T.-K.L. Wong, 2015 "Volatility harvesting in theory and practice." *The Journal of Wealth Management*, vol. 18, no. 3: pp 89–100.

Bullen, P.S, 2009, "Accentuate the negative." *Mathematica Bohemica*, vol. 134, no. 4: pp 427–446.

Campbell, J.Y., A.W. Lo, and A. Craig MacKinlay, 1996, *The Econometrics of Financial Markets*, Princeton University Press. Princeton, New Jersey.

Chambers, R.D. and J.S. Zdanowicz, 2014, "The limitations of diversification returns." *Journal of Portfolio Management*, vol. 40, no. 4: pp 65–76.

Dempster, M.A.H., I.V. Evstigneev, and K.R. Schenk-Hoppé, 2007, "Volatility induced financial growth." *Quantitative Finance*, vol. 7, no. 2: pp 151–160.

Dichtl, H., W. Drobetz, and M. Wambach, 2014, "Where is the value added of rebalancing? A systematic comparison of alternative rebalancing strategies." *Financial Markets and Portfolio Management*, vol. 28, no. 3: pp 209–231.

Fernholz, R. and C. Maguire, 2007, "The statistics of statistical arbitrage." *Financial Analysts Journal*, vol. 63, no. 5: pp 46–52.

Granger, N., D. Greenig, C.R. Harvey, S. Rattray, and D. Zou, "Rebalancing Risk" (October 3, 2014). Available at SSRN: https://ssrn.com/abstract=2488552 or http://dx.doi.org/10.2139/ssrn.2488552

Hallerbach, W.G., 2014, "Disentangling rebalancing return." *Journal of Asset Management*, vol. 15, no. 5: pp 301–316.

Hamilton, J.D., 1994, *Time Series Analysis*, Princeton University Press. Princeton, New Jersey

Hayley, S., K. Cuthbertson, N. Motson, and D. Nitzsche, "Diversification returns, rebalancing returns and volatility pumping" (January 14, 2015). Available at SSRN: https://ssrn.com/abstract=2311240 or http://dx.doi.org/10.2139/s srn.2311240

Lo, A.W. and A.C. Mackinlay, 1990, "When are contrarian profits due to stock market overreaction?" *The Review of Financial Studies*, vol. 3, no. 2: pp 175–205.

Mindlin, D., 2011, "On the relationship between arithmetic and geometric returns." CDI Advisors, LLC.

Perold, A.F. and W.F. Sharpe, 1988, "Dynamic strategies for asset allocation." *Financial Analysts Journal*, vol. 44, no. 1: pp 16–27.

Qian, E.E., 2011, "Risk parity and diversification." *The Journal of Investing*, vol. 20, no. 1: pp 119–127.

Qian, E. E., 2012, "Diversification return and leveraged portfolios." *The Journal of Portfolio Management*, vol. 38, no. 4: pp 14–25.

Qian, E. E., "To rebalance or not to rebalance: A statistical comparison of terminal wealth of fixed-weight and buy-and-hold portfolios" (January 26, 2014). Available at SSRN: https://ssrn.com/abstract=2402679 or http://dx.doi.org /10.2139/ssrn.2402679

Rabault, G., "Harvesting excess volatility" (August 2017). Available at SSRN: https://ssrn.com/abstract=3146917 or http://dx.doi.org/10.2139/ssrn.3146917

Rulik, K., 2013, "Can the rebalancing bonus enhance beta return?" *Journal of Indexes*. Available at ETF.com: https://www.etf.com/publications/journal ofindexes/joi-articles/20200-can-the-rebalancing-bonus-enhance-beta-re turn.html?

Willenbrock, S., 2011, "Diversification return, portfolio rebalancing, and the commodity return puzzle." *Financial Analyst Journal*, vol. 67, no. 4: pp 42–49.

Index

Printed in the United States
by Baker & Taylor Publisher Services